D0341423

Neptune's Ark

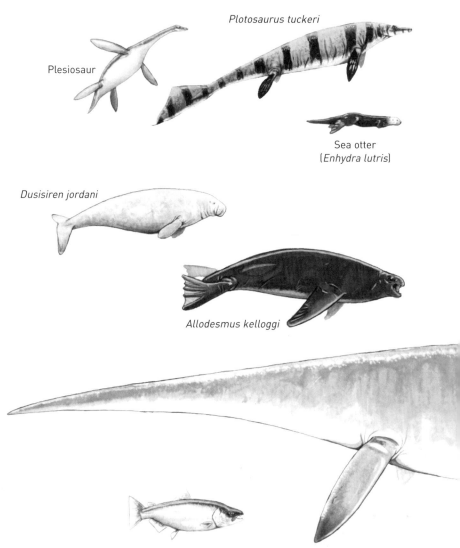

Plesiosaur

Plotosaurus tuckeri

Sea otter
(Enhydra lutris)

Dusisiren jordani

Allodesmus kelloggi

Smilodonichthys rastrosus

Hesperornis regalis

Aetiocetus cotylalveus

Orca
(*Orcinus orca*)

Shonisaurus sikanniensis

Human
(*Homo sapiens*)

Neptune's Ark

FROM ICHTHYOSAURS TO ORCAS

David Rains Wallace

Illustrations by Ken Kirkland

UNIVERSITY OF CALIFORNIA PRESS

BERKELEY   LOS ANGELES   LONDON

*The publisher gratefully acknowledges the generous
contribution to this book provided by the General
Endowment Fund of the University of California
Press Foundation.*

University of California Press, one of the most
disinguished university presses in the United States,
enriches lives around the world by advancing scholarship
in the humanities, social sciences, and natural sciences.
Its activities are supported by the UC Press Foundation
and by philanthropic contributions from individuals
and institutions. For more information, visit
www.ucpress.edu.

University of California Press
Berkeley and Los Angeles, California

University of California Press, Ltd.
London, England

Library of Congress Cataloging-in-Publication Data

Wallace, David Rains, 1945–.
    Neptune's ark : from ichthyosaurs to orcas / David
Rains Wallace ; illustrations by Ken Kirkland.
        p.   cm.
    Includes bibliographical references and index.
    ISBN 978-0-520-24322-4 (cloth, alk. paper)
    1. Vertebrates—Pacific Coast (U.S.)—History.
2. Marine animals—Pacific Coast (U.S.)—History.
I. Title.
QL606.52.U6W35   2007
596'.0979—dc22   2006035806

Manufactured in the United States of America

16   15   14   13   12   11   10   09   08   07
10   9   8   7   6   5   4   3   2   1
The paper used in this publication meets the minimum
requirements of ANSI/NISO z39.48–1992 (R 1997)
(*Permanence of Paper*).

To the memory of T. H. Watkins

We do not associate the idea of antiquity with the ocean, nor wonder how it looked a thousand years ago, as we do of the land, for it was equally wild and unfathomable always. The Indians have left no tracks on its surface, but it is the same to the civilized man and the savage. The aspect of the shore only has changed. The ocean is a wilderness reaching round the globe, wilder than a Bengal jungle, and fuller of monsters, washing the very wharves of our cities and the gardens of our sea-side residences. Serpents, bears, hyenas, tigers rapidly vanish as civilization advances, but the most populous and civilized city cannot scare a shark from its wharves. It is no further advanced than Singapore, with its tigers, in this respect. The Boston papers had never told me that there were seals in the harbor. I had associated these with the Esquimaux and other outlandish people. Yet from the parlor windows all along the coast you see families of them sporting on the flats. They were as strange to me as the mermen would be. Ladies who never walk in the woods, sail over the sea. Why, it is to have the experience of Noah, —to realize the deluge. Every vessel is an ark.

—*Thoreau*, Cape Cod

CONTENTS

ILLUSTRATIONS

ACKNOWLEDGMENTS

I wish to thank a number of individuals and institutions that helped with this book. Blake Edgar, science acquisitions editor at University of California Press, was supportive of a proposal from a nonexpert in a crowded and competitive field. Blake, Gary Snyder, David Phillips, George Schaller, and Clayton E. Ray kindly provided references. Clayton, curator emeritus at the Smithsonian's National Museum of Natural History, assisted with a trip to the museum that was crucial to my research. There he, and collections manager David Bohaska and librarian Martha Rosen, were gracious hosts, answered many questions, showed me fossils and other materials, and generally made my visit rewarding. They put me in touch with Daryl Domning, professor of anatomy at Howard University, and Storrs Olson, curator of ornithology at the NMNH, who talked to me about their areas of expertise. Ellen Allers and Keith Gorman of the Smithsonian Institution Archives were very accommodating in making reference materials available before and during my visit. Also helpful in answering questions and supplying research materials were Larry Barnes of the Los Angeles County Museum, Kevin Padian of the UC Berkeley Museum of Paleontology, Doug Kennett of the University of Oregon, and Carola de Rooy of Point Reyes National Seashore. My thanks to the Doe, Bancroft, Biosciences, Geology, Periodicals, and Anthropology libraries at the University of California and the library of Southern Oregon University in Ashland for the use of their collections.

Steller's Sea Ape

The coast trail at Point Reyes National Seashore follows one of the few roadless stretches on the California coast, so it is a good place to notice things. Walking on it one day, I heard a ludicrous racket from the water, like parrots gargling. When I moved to the clifftop, I saw hundreds of little penguin-like seabirds, common murres, on a large rock just offshore, screaming their heads off. I had seen the rock before, scattered with silent pelicans and cormorants and the occasional vociferous gull, but never packed with screaming murres. It was exhilarating to encounter such a mob in what had seemed an inert, if scenic, bit of coast.

That place on Drake's Bay, called Point Resistance, stuck in my mind, and I often visited it afterward. I found that the murres didn't make their racket only in late spring and summer, when they packed the rock. In late winter and early spring, they packed the surrounding water, rafts of birds that paddled around the island in a kind of marine cotillion. But, despite their ardent clamor, I saw little evidence of why they frequented the rock. I assumed it was for breeding, but I couldn't see eggs or chicks even with binoculars, except a few times when ravens swooped down, scattered the murres, and carried greenish spheres away. And the murres weren't always there. Packed with them one day, rock and water might be empty a week later.

It was as though Point Resistance sometimes awakened from a prevailing slumber, but only partly, or as though the murres were phantoms from the time before gill nets and oil spills, when such mobs perhaps did breed on every rock along the coast. The place seemed under a spell. I felt tranced myself, watching the endless cycle of kelp waving in blue depths, silt billowing in green shallows, and surf breaking on beige sand.

It was like shores I'd seen from Mexico to Alaska, a crescent of beach, cliff, and headland backed by peaks and ridges. Such places often seem inert, although they are the opposite geologically. North America's Pacific coast is one of the more dynamic. Point Resistance demonstrates this. The upper cliff contains level sand-and-gravel sediments pushed up from the water by tectonic movements in recent millenia, the lower crazily tilted and twisted metamorphosed strata from a much older time, an "unconformity" manifesting a very long hiatus. The surf has carved ragged caves and arches in the hard lower strata, belying the Pacific's name. Yet this evidence of perpetual violence fails to dispel a sense of repose. The surf often lulled me asleep, to awaken with a start at the thought of rolling off the cliff.

Of course, much waking life is scattered around Point Resistance. Gulls and vultures patrol the cliff, and peregrine falcons perch on the knife-backed arch that ends the point. Sometimes harbor porpoises pass offshore, briefly showing small dorsal fins. A harbor seal usually bobs in the shallows or upends itself to dive, its underwater pursuits obscured by the wind-ruffled surface. Once an osprey zoomed past me, did a startling dip and roll that almost crashed it into the cliff, then swooped to the surf's edge simply to snatch a bit of kelp.

In winter, many seabirds rest in the surf—loons, scoter ducks, cormorants, mergansers—and black oystercatchers swoop around the rock, even noisier than the murres. Groups of male California sea lions swim southward, flustering the murre cotillions as they rise to breathe, and a few gray whales pass toward their Baja breeding grounds. In spring, pigeon guillemots fly into cliff holes, cormorants carry kelp to the rock, and the sky may contain half a dozen ospreys at once. Numerous gray whales swim north toward Bering Sea feeding grounds, hugging the shore to protect their calves from orcas. In summer, brown pelicans arrive to do their pterosaur impression, and the breeze reeks of guano from the newly white-washed rock. In fall, willets, marbled godwits, and sanderlings flock around the beach like drifting sand.

But such happenings retain a tranced quality. Although dozens of whales may pass in a few hours in April and May, they rise, blow, and submerge so quickly that their presence seems spectral. Sea lion and harbor porpoise sightings are as transitory. I've rarely seen an osprey or a harbor seal catch a fish at Point Resistance, and even the cormorants and loons seldom do, as though the need to eat was suspended in the place. One afternoon, I realized that a seal that kept rising and sinking at a spot in the shelter of the arch was sleeping in what must have been a restful marine bedroom.

Still, I feel less confidence in the point's apparent languor than I would about a similar inland spot. More odd things seem to happen than on land. On many days when I see harbor porpoises, countless half-dollar-sized *Velella* hydroids, jellyfish relatives with translucent white sails and iridescent violet-blue bodies litter the beaches—doubtless a coincidence, but odd. Unidentified cetaceans pass offshore of the murre rock, showing taller dorsal fins than the porpoises. Unexplained dark shapes and splashes occur. Point Resistance may be just a bedroom of the Pacific, but so many things lurk in that Neptunian mansion that there is no knowing what might slip in. When the air is clear, a notorious haunt of monsters looms to the west, conical sea mounts brooding on the horizon like the lost world of *King Kong*.

On a trip to the mounts, prosaically named the Farallones, the cliff islands, I've seen more kinds of monsters than anywhere else. Past the Golden Gate, swarms of barrel-sized lion's mane jellyfish attracted two eight-foot leatherback turtles, which poked heads tinted red and purple with marine epiphytes from the water. Hot-tub-sized sunfish lay sideways on the surface, gazing up with saucer eyes. Near the islands, humpback and gray whales flourished their flukes, perhaps diving for small fish and amphipods. On the islands, pinnipeds festooned rocks and beaches—silver gray elephant seals, mottled harbor seals, golden brown Steller's sea lions, chocolate brown California sea lions. Northern fur seals may have been there too, but I couldn't discern any in the crowd. In the open sea beyond, sea lions and black-and-white Dall's porpoises followed the boat, the lions diving back and forth underneath it, the porpoises surfing the bow wave.

That was a normal Farallones day in late summer, and there is always a chance of seeing more monstrous giants. In 1999, passengers on a Farallon tour boat saw an orca attack a great white shark and carry it in its mouth at the surface for several minutes, a thirty-foot monster brandishing a thirteen-foot one. Blue whales, the largest animals ever, appear more often since the 1986 whaling moratorium. In 1880, a steam launch shot a "patented bomb lance" into the tail of one in Drake's Bay. According to an observer, it towed the ten-ton boat "for three or four hours, with from twenty to forty fathoms of whale line, sometimes at a rate of ten miles an hour, although we were frequently backing under a full head of steam." Then it broke the line and swam away.

And all these giants live extraordinary lives. Gray whales make the longest annual migration of any living mammal, traveling some 12,000 miles each year between breeding and feeding grounds. The grays follow the coast, but it is less clear how white sharks or humpback and blue whales navigate throughout the world's oceans, or how elephant seals, which migrate far offshore to feed, return to small breeding islands like the Farallones. Humpback whales make a variety of strangely beautiful sounds, which may have been the source of classical antiquity's Siren legends. Orcas swallow seals in a gulp and have been found with over a dozen in their stomachs. Elephant seals dive a mile deep and stay down over an hour to feed on squid, skates, and ratfish. Leatherback turtles dive a half mile deep in pursuit of jellyfish.

The coast's unpredictable mix of vacancy and monstrosity has a peculiar mental resonance. How does such gigantism arise from lifeless water, sand, and sky? The most memorable dream I've ever had was of standing on a lonely beach as sea creatures came ashore. There was distant music like the echoes in a conch, and a sense of exaltation so great that I can still recall it, forty years later.

The dream creatures included seals, dolphins, seabirds, and turtles. The main figures were mysterious, however. They had long-eared, maned heads like land animals, but serpentine bodies and finned tails like fish, unlike anything I've seen in the real ocean. Of course, fantastic beings are common in dreams, and those in mine were very like the fish-tailed horses and other merbeasts of myth. But they gave the dream much of its power. They seemed to link sea and land miraculously. And what is the source of such myths?

One of history's most mythic voyages was Vitus Bering's 1741 Russian expedition to Alaska, which, although this now seems surprising, was still *terra incognita.* Alaska defied discovery even then. The expedition was able only to briefly explore a few foggy offshore islands before storms and scurvy drove it back along the Aleutians. A sense of unreality overhangs it, and an encounter that remains unexplained epitomizes this. On August 10, 1741, when Bering's flagship, the *Saint Peter,* was becalmed at sunset south of Kodiak Island, an expedition naturalist, Georg Wilhelm Steller, described a "very unusual and new animal," like the ones in my dream:

> The animal was about two ells [six feet] long. The head was like a dog's head, the ears pointed and erect, and on the upper and lower lips on both sides whiskers hung down which made him look almost like a Chinaman. The eyes were large. The body was longish, round and fat, but gradually

became thinner toward the tail; the skin was covered thickly with hair, gray on the back, reddish white on the belly, but in the water it seemed to be entirely red and cow-colored. The tail, which was equipped with fins, was divided in two parts, the upper fin being two times as long as the lower one, just like on the sharks.

However, I was not a little surprised that I could perceive neither forefeet as in marine amphibians nor fins in their place.

The creature enchanted Steller and his shipmates:

For more than two hours it stayed with our ship, looking at us, one after the other, with admiration. It now and then came closer and often so close it could have been touched with a pole. Then, as soon as we moved, it retired further away.

It raised itself out of the water up to one third of its length, like a human being, and often remained in this position for several minutes.

After it had observed us for almost half an hour, it shot like an arrow under our ship and came up again on the other side, but passed under the ship again to reappear in its first position. It repeated this maneuver back and forth over thirty times.

Now, when this animal spotted a large American seaweed, three to four fathoms long, which at the bottom was hollowed out like a bottle and from there to the outermost end became gradually more pointed like a phial, it shot toward it like an arrow, grabbed it in its mouth, and swam with it toward our vessel, and did such juggling tricks that one could not have asked for anything more comical from a monkey. Now and then it bit off a piece and ate it.

This description of a "sea ape" has baffled biologists ever since, not so much for its oddness, common enough in early exploration accounts, as for its detailed sobriety. Georg Wilhelm Steller was the least whimsical of early naturalists, a scrupulous worker whose punctiliousness made him unpopular. In a treatise on marine mammals written during the expedition, he scorned "imperfect histories . . . swarming with fables and false theories . . . in which the writers of natural history saw only through a lattice what they might have seen with their own eyes." Science has verified the existence of every other animal that he mentioned. He was the first naturalist to describe scientifically the sea otter, northern fur seal, and Steller's sea lion. He was the only one ever to describe a creature as odd as the "very unusual and unknown" one of August 10.

By November, Bering's ship was hopelessly lost, and a storm stranded it on the Aleutians' western end. Steller hoped they'd reached their Siberian base, the Kamchatka Peninsula. Walking on the shore when the weather cleared, however, he was astonished to find the shallows filled with "many manatees, which I had never seen before—nor could I even know what kind of animal it was since half of it was constantly under water." The expedition had stumbled not on Kamchatka, but on an uninhabited island, now called Bering Island, that was the last refuge of the northern sea cow, a sirenian like the manatee and dugong. Since known sirenians inhabited only warm seas, Steller's surprise at finding a subarctic one was extreme, not least in that it was three times bigger. The stomach of a female he dissected was five feet wide, six feet long, and so stuffed with kelp that four men could barely budge it with ropes.

The more he saw of the sea cows, the more Steller's wonder grew:

> The largest of these animals is four or five fathoms [thirty feet] long and three and a half fathoms thick around the region of the navel, where they are thickest. Down to the navel it is comparable to a land animal; from there to its tail, a fish. . . . In place of teeth, it has in its mouth two broad bones, one of which is affixed above to the palate, the other on the inside of the lower jaw. Both are furnished with many crooked furrows and raised ridges with which it crunches seaweed as its customary food. . . . Below, on the chest, there are two strange things to be seen. First, the feet, consisting of two joints, have outermost ends rather like a horse's hoof. Underneath, these are furnished with many short and densely set bristles like a scrub brush, and I am not prepared to say whether to call them hands or feet. . . . With these forefeet, it swims ahead, beats the seaweed off the rocks in the bottom, and when, lying on its back, it gets ready for the Venus game, it embraces the other with these, as with arms. . . . The second curiosity is found under these forefeet, namely the breasts, provided with black, wrinkled, two inch long teats, at whose outermost ends innumerable milk ducts open.

Discovering such a beast would have seemed to vindicate the earlier sighting. To be sure, Steller was stranded on Bering Island until late 1842 and had months to study the sea cow, about which familiarity bred matter-of-factness. The beasts were foolishly gregarious, crowding around wounded companions, allowing more to be killed. Their meat was as good as beef, their rendered fat so tasty that men could drink bowls without nausea. "These animals are found at all times of the year everywhere around this island in vast

Figure 1. Steller's sea cow, *Hydrodamalis gigas,* is related to living dugongs. The last of several sirenian species to inhabit the west coast, and the only one adapted to subarctic waters, it was hunted to extinction by 1768, less than thirty years after its discovery.

numbers so that the coast of Kamchatka could continually supply itself plentifully from them with both fat and meat," Steller enthused—an uncharacteristic, and unprophetic, speculation. Fur hunters, supplying nobody but themselves, would exterminate Steller's sea cow by 1768.

Steller tried to be matter-of-fact regarding the sea ape but only managed to increase its mystery:

> When I had observed it for a long time, I had a gun loaded and fired
> at this animal, intending to get possession of it and make an accurate
> description. But the shot missed. Although it was somewhat frightened,
> it reappeared right away and approached our ship gradually. But when
> another shot at it was in vain, or perhaps only slightly wounded it, it
> retreated into the sea and did not come back. However, it was seen at
> various times in different parts of the sea.

Historians have tried ever since to make sense of this fantastic report from an otherwise reliable observer. Steller's scholarly biographer, Leonhard Stejneger, concluded in the 1930s that the sea ape was a northern fur seal, a species that, Stejneger thought, Steller had not seen at that point. "Most of the peculiarities of behavior described are quite characteristic of such an animal when in a playful mood and not frightened," Stejneger wrote. "Certain of the bodily characters described are also those of the fur seal, such as the dog-like face, the large eyes, the long overhanging whiskers, the pointed ears." Victor B. Scheffer, a marine mammal biologist, concurred in 1970 that the creature was a fur seal of which Steller had failed to see the front flippers.

It may not be that simple. Although Steller's original expedition journal has disappeared, an old handwritten copy found by a Stanford professor in the Petrograd Academy of Sciences includes a 1741 passage that undermines Stejneger's explanation: "On August 4, when sailing to the south, we saw finally between south and west about two or three miles from us many high, large, and wooded islands, so that we were surrounded by land. . . . During this time we spent close to land, we constantly saw large numbers of fur seals, other seals, sea otters, sea lions and porpoises." If Steller had been identifying "large numbers of fur seals" at least since August 4, he would have recognized one that he watched for two hours a week later. His kelp-eating observation also undermines a fur seal explanation, since fur seals are carnivores like other pinnipeds. Corey Ford, author of a 1966 book about the expedition, maintained that "as trained and exact a scientist as Steller could scarcely have confused it with an otter or seal."

But then, as the Petrograd copy's editors plaintively ask, "What did Steller see?" Ford thought that "the simplest explanation is that the 'sea monkey' actually existed, and that Steller saw it for the first and last time before it became extinct like the northern sea cow." In a 1994 book on sea monsters, Richard Ellis, an author and illustrator, guessed that Steller saw a Pacific version of the "sea mink," a historically extinct mammal of the North Atlantic coast, which he described as "about twice as large as the Eastern mink (*M. vison*)" with "coarser, reddish-brown fur."

Ellis's idea failed to explain the weirdness of Steller's August 10 sighting, however. Thomas Pennant, an eighteenth-century naturalist, called the sea mink a "strange animal" but said nothing about missing forelimbs or fish-tails. According to an eyewitness description Pennant quoted from a "Mr. Phipps" in Newfoundland, it was "of a shining black: bigger than a fox, shaped like an *Italian* greyhound; legs long; tail long and taper. One gentle-men saw five sitting on a rock with their young, at the mouth of a river; often leapt in and dived, and brought up trouts, which they gave to their young."

Ellis complicated his idea with reports of creatures that sound like larger versions of the sea ape, perhaps aloof seniors of Steller's playful six-footer. He quoted two Canadian zoologists, Paul LeBlond and John Silbert, who since 1969 had collected many "observations of large unidentified animals" with horselike or sheeplike heads, serpentine, furry bodies, and fishlike tails in British Columbia waters. Querying marinas, lighthouses, fishing clubs, and newspapers, they had "isolated 23 sightings which could not definitely or even speculatively be accounted for by animals known to science." Ellis included a fuzzy 1937 photo of an elongated object allegedly found in a sperm whale's stomach. Propped on a dock, it might have been the skull, spine, and tail of an equine-headed, serpentine creature.

In 1995, LeBlond, then director of the University of British Columbia's Program of Earth and Ocean Sciences, published a book on the creatures, which he named *Cadborosaurus*. He and another zoologist, Edward L. Bousfield, listed hundreds of descriptions going back to the nineteenth century and cited cryptobiology's doyen, Bernard Heuvelmans, on its place in the bestiary of the unknown: "Heuvelmans, in his voluminous review of world-wide serpents, grouped together all the 'elongated sea animals of large size characterized by a medium length neck, a mane, a horse's head and large eyes' under the label 'Merhorse' and offered a sketch . . . in which some of Caddy's features, especially its head and neck, can be recognized."

LeBlond's book reproduced clearer photos of the 1937 carcass Ellis had shown, describing it as "essentially intact and easily distinguishable from any marine animal previously known to the whaling station men." According to an eyewitness, "it had a horselike head with large limpid eyes and a tuft of stiff whiskers on each cheek. Its long slender body was covered by a fur-like material, with the exception of its back, where spiked horny plates overlapped each other." The station's annual report described it as "about 10 feet long, having a head similar to a large dog's, animal-like vertebrae, and a tail resembling a single blade of gill bone, as found in a whale's jaws."

People have reported elongated, furred creatures with horse heads and shark tails as far south as California. That doesn't prove they exist, of course. The carcass photographed in 1937 has disappeared, and no other specimen is known, which is not true of Steller's sea cow. Although no complete specimen exists, museums have historic sea cow bones from Bering Island as well as prehistoric fossils of the genus *Hydrodamalis* and other west coast sirenians. But no bones of fish-tailed, horse-headed animals are in collections, and the genus name *Cadborosaurus* will remain scientifically invalid until a specimen turns up.

The sea ape may have been a hoax, foisted on Steller by mischievous shipmates. He may have concocted it himself to mock the "fables" he deplored, or a rival may have done so to damage his reputation. Peter Simon Pallas, a Russian naturalist who published Steller's journal in the late eighteenth century, rearranged and rewrote parts of it. Someone could have tampered with the Petrograd copy. But there is no way of knowing this without more evidence, and, interestingly, the sea ape story has not had the discrediting effect a hoaxer might have intended. If anything, it has enhanced both the mystique of early exploration and Steller's fame as an unprejudiced and exact observer. Therein seems to lie its significance until a specimen turns up, or until people stop reporting "mer-horses."

Whether or not such creatures have an objective existence, they have a subjective one. A longing for mysterious wonders is universal, and where better to seek them than the place where the land, dulled by use, meets the ocean? John Steinbeck and his marine biologist friend, Ed Ricketts, wrote of their own mythic voyage to Baja California: "There is some quality in man that makes him people the ocean with monsters. . . . An ocean without its unnamed monsters would be like a completely dreamless sleep." Far from vitiating the grandeur of Bering's voyage to what was then the end of the Earth, Steller's improbable creature seems to enhance it.

Since the voyage, science has affirmed a wondering view of the coast with its revelations of the marvelously diverse abundance of marine life, not only now, but during its long prehistoric past. Fossils of extinct marine creatures as strange as the sea ape have come to light—whale-sized reptiles; ursine clam-eaters; elephantine seagrass-eaters; "toothed" birds with twenty-foot wingspans; giant, saber-toothed salmon. There are even marine sloth fossils, albeit only from South America's west coast so far.

Evolving marine diversity has indeed transformed the land, as reflected in my dream. And nowhere else have hopes for new human relationships with the sea's strange old life run higher than on the North American west coast that Steller helped to discover. Those hopes, growing over the past century, have helped to allow a marine diversity to survive and even recover some abundance in places such as Drake's Bay and the Farallones, only a few miles from the world's most explosive cities.

Of course, there is another side to the sea-land transition. Steller noted it just after the August 4 marine mammal sightings that preceded the sea ape: "I frequently observed that as often as these sea animals allowed themselves to be seen, even in greatest calm, thereafter the weather changed, and the more frequently they appeared and the more movement they made, the more furious the storms were." The sea ape's enchantment was a brief interlude in a welter of disasters that left a quarter of the *Saint Peter*'s crew dead, including Vitus Bering, before the expedition finally limped back to Kamchatka. Steller worked steadfastly throughout the disasters, but he never recovered from the ordeal and died a few years later, still in his thirties.

From this other side, the seacoast can seem the opposite of my dream, a

place where life's limits instead of its wonders manifest themselves, not only ·
practically, as with storms, scurvy, and shipwrecks, but mythically. The
mythic coast is not only where life comes ashore, but where it passes away.
The Coast Miwok who lived north of San Francisco Bay thought so. In
1931, a Miwok man named Tom Smith told anthropologist Isabel Kelly:
"The dead leapt into the ocean at Point Reyes and followed 'a kind of string
leading west through the surf,' to a road beyond the breakers. This took
them farther west to the setting sun, and there they remained with Coyote
in the afterworld." Such myths run deep. At her husband's funeral recep-
tion overlooking Monterey Bay, my sister, a PhD in speech pathology, saw
a passing porpoise pod and said: "Well, there he goes."

Science also has affirmed a view of the seacoast as a place where life meets
limits. It is the "lower" forms of life that have transformed the Earth by
coming ashore—the invertebrate and fish ancestors of land animals. Paleontology shows that "higher" land-evolved vertebrates—now called tetrapods—have been returning to the sea's abundance almost since adapting to
land, but shows little movement in the other direction. Fossil equivalents
of my dream of sea creatures transforming into land ones are almost
unknown. For tetrapods, the evolutionary movement has been toward
ever-increasing adaptation to marine life, leading to highly specialized
forms and eventually, often for unclear reasons, to extinction.

When I encounter marine creatures come ashore in waking life, it is usually as injured waifs or rotting carcasses. The beaches around Point Resistance are sun-whitened sepulchers of dead things thrown up by winter
storms, buried by sand, and sometimes exhumed by new storms. Even
when creatures come ashore voluntarily, it often is deadly, as when whales
strand themselves, orcas surge out of surf to grab unwary prey on a beach,
or dolphins trap fish by driving schools onto riverbanks.

Our dual mythos of the coast as a place of life to be embraced and a place
of death to be resisted engenders a tense ambiguity. Often, particularly since
humans found places like Bering Island, but long before that, the tension
has driven us to love the coast to death by assaulting an abundance that
seems mindlessly profligate as well as marvelous, that seems to invite wan-
ton destruction. The tension has never been greater than it is now, when
dreams of transcendent revelation—of superior cetacean intelligence and
new partnerships between human and marine life—conflict with relentless
commercial demands on every maritime resource.

The tension is nowhere higher than along North America's Pacific coast,
in places like Point Resistance. As whales, porpoises, and sea lions pass, San

Francisco's smog browns the southern horizon, jumbo tankers loom offshore, and a bewildering array of trash litters the beach, from ship timbers and fishnet floats to hypodermic needles and detergent bottles. Uncontrolled economic growth makes natural abundance seem precarious, a fragile membrane stretched between civilization's petrochemical explosions and the galaxy's thermonuclear blasts.

Still, North America's west coast is durable as well as fragile. It has seen many mass extinctions and other disasters, going back to life's emergence onto land and before. If there is such a thing as an "ur-coast" on the planet, the thousands of miles north and south of Point Resistance might come as close to it as any. Today's east coast was deep in the Laurasian supercontinent before tectonic spreading centers began to open the Atlantic Ocean some 200 million years ago, and it had lain in the interior of even older land masses. North America's western margin has bordered oceans at least since hard-skeletoned animals began to dominate the sea 550 million years ago.

The margin was farther east for most of that time. Western North America has only gradually reached its present size through accretion of island arcs and other "exotic terranes" rafted onto it by crustal plate movements. Even when the continent had grown to what is now Nevada, shallow seas flooded its interior for many millions of years, causing it to have more than one "west coast." The coast did not reach anything like its present form and location until the Eocene epoch, fifty million years ago. But tectonic turmoil assured that an interface of land and sea always existed, and the fossil record reflects this. There is a great antiquity of coastal interaction here.

That antiquity is not only prehistoric. The west coast was the last in the Northern Hemisphere to be discovered for commercial exploitation, and it had animals found nowhere else—Steller's sea cows; animals not found in other northern seas—sea otters, sea lions, fur seals, and elephant seals; and animals exterminated elsewhere—gray whales. That exploitation lag did much to allow today's precariously surviving biodiversity, making the west coast a kind of "Neptune's ark" against the fiery catastrophe biblically predicted to succeed the watery one.

The west coast's evolutionary continuum has its confusions and hiatuses. Subduction and other tectonic violence have prevailed along its colliding oceanic and continental plates, cooking many sediments into hard metamorphic rocks, so it largely lacks the east coast's continuous swathes of fossil-bearing marine strata. Coastal fossils tend to be scattered in inconvenient places like cliffs and surf-struck beaches. Tectonic violence may have expunged some ancient creatures that survive in the Atlantic, like *Limulus,* the "horseshoe crab," a scorpion relative little changed from 200-million-year-old fossils. *Limulus* was laying its eggs along Atlantic beaches as that ocean formed, and it survives on Asia's Pacific coast but is gone from North America's (although the *San Francisco Evening Bulletin* reported the capture of one off the Farallones on May 29, 1886).

The interaction of sea and land still has left enough evidence to tell a story, albeit a sometimes hectic one. And maybe that is more appropriate to the coast of gold rush and Hollywood than a more sober narrative. Not that sensational sea monsters haven't been reported elsewhere. But, for all its bafflement, the story of Steller's sea ape is unsurpassed for drama and atmosphere. And drama continued in the discovery of the west coast's prehistory. In the nineteenth century, it became an arena of history's greatest paleontological feud. In the twentieth, it remained a frontier of paleontological exploration, and some of that was as mythic and tragic as Steller's.

Steller's sea ape seems emblematic of such dramas, because it concerns not only the sea and its creatures but the baffling relationship between them and the mind. And I will tell it not as a marine authority but as a fascinated but uneasy land creature. We know little enough about our own evolutionary niche, so I've felt that exploring it with the land-oriented senses we have was enough for a financially and technologically challenged observer. But the sea is in my blood, as in all animals' (containing the same salts, although in more dilute solution), so I will try to open a porthole on the greatest living realm by telling a little about the past billion years on one of its continental margins.

ONE

Reefs in the Desert

The fact that the sea is in our blood is merely the most intimate evidence that life began in the ocean. There are many others, including the fact that most life is still there. The evidence doesn't explain how or why life began there. Origin theories are legion, and perhaps as unverifiable as Steller's sea ape. The idea that the first organisms evolved from complex molecules in the highly energetic environment of seafloor volcanic vents is popular now. It will be hard to test paleontologically because such environments don't fossilize well. It seems likely, anyway, that life occupied the sea very early.

J. William Schopf, an authority on early life, thinks an Australian rock formation called the Apex Chert contains fossils of filamentous bacteria and other prokaryotes, cells without nuclei, that may be more than 3,465 million years old. He thinks the organisms formed mats in the shallows surrounding early continents, which were like large islands in the vast primal oceans. "The scene was dominated by broad shallow seaways into which volcanic lavas erupted," Schopf writes of that environment. "Scattered volcanic islands were fringed by river gravels, sandy inlets, mudflats, and occasional evaporitic lagoons."

It is hard to tell if such unimaginably ancient structures are really the remains of life, but similar organisms definitely inhabited such environments very long ago, and for a very long time. Fossils from Montana's 1,300-million-year-old Belt Supergroup are of filamentous cyanobacteria growing in clumps called stromatolites. They still live in them, on North America's west coast among other places. Salty lagoons in Baja California, where more recent algae-eaters like snails can't survive, support stromatolites that resemble fossil ones, although the constituent species are different.

Fossils suggest that life underwent much of its further evolution from prokaryotic to eukaryotic organisms, cells with nuclei, and then to multicellular organisms in the seas. Fossils of unicellular eukaryotes are known from 1,800-million-year-old marine rocks. In western North America, the 850-million-year-old Kwagunt Formation from Arizona has yielded a "vase-shaped" protozoan like something one might see in plankton today.

Charles D. Walcott, a paleontologist with the U.S. Geological Survey, discovered some of the first really ancient fossils in Arizona. Exploring the Grand Canyon in the 1880s, he found algal reef structures from Precambrian times, over 550 million years ago. In 1899, again in the Grand Canyon, Walcott found tiny blackish disks, which he named *Chuaria* after the rock formation. He interpreted them as fossils of shelled animals, but they later turned out to be giant eukaryotic algal cells.

It took longer for paleontologists to find very early multicellular organisms. In 1946, strange saucer-shaped fossils turned up in 560-million-year-old rocks in an abandoned mine named Ediacara (an aboriginal word for a spring or seep) in southern Australia. Although unimpressive, they were too large and complex to be unicellular. As more fossils of similar structure and age emerged from places like Namibia and northern Russia, paleontologists realized that a diverse array of small and soft-bodied, but many-celled, organisms quite unlike most living ones had evolved by at least 800 million years ago.

More recent discoveries have shown that this "Ediacaran" ecosystem also occupied the gradually forming western margin of North America. In 1996, Mark McMenamin, a geologist, reported the discovery of 600-million-year-old "body and trace fossils" of "sediment-dwelling animals" in Mexico's northwestern state of Sonora. The body fossils were "headless, tailless, and appendageless." One of the the fossils, named *Cyclomedusa,* was shaped somewhat like a fried egg and may have dwelt on the ocean bottom like a sea anemone. Another, conical one, *Sekwia,* may have been the "semi-rigid basal attachment structure" of another bottom-dwelling Ediacaran. The siltstones and sandstones in which the fossils were embedded suggested that they had lived in shallow waters, probably "on or near the continental shelves of the North American craton."

Ediacaran fossils similar in age to the Sonoran ones have been found in northwest Canada's MacKenzie Mountains, implying that, even then, coastal environments extended thousands of miles along the continent's edge. Somewhat younger Ediacaran fossils have since turned up in the deserts of southwest Nevada and southeast California, and in northern California's Klamath Mountains. The remains of disklike, tubular, or frondlike organisms, they have similarities to Brazilian and Namibian fossils, suggesting that this continent lay across a narrow, newly forming ocean from what are now parts of Africa and South America.

Nobody is sure what the Ediacaran organisms were or how they lived—whether they had mouths and guts, absorbed food directly from water, or had symbiotic relationships with photosynthetic organisms. Various living organisms feed in all these ways, and it is possible that the Ediacarans were the ancestors at least of some of them, but nobody is sure about that either. The fossil record, anyway, shows that the Ediacarans and ancient versions of living groups inhabited Precambrian and Cambrian seas together for many millions of years before the Ediacarans disappeared.

The southwestern Ediacaran fossils confirm an impression of dessicated marine antiquity that I got from the California desert when I first went there in the 1980s. Hiking in the eastern Mojave, I thought I saw evidence of an abundance of small carnivores—kit foxes, ringtails, and the like—assuming that black cylinders on most rocks were their scats. Then I realized that the cylinders were not mammal scats, but Paleozoic invertebrate fossils.

I also realized, looking at a mountain east of Death Valley, that what I was seeing was not just a pile of minerals but the petrified remains of an enormous reef formed by archaic organisms. To look closely at the mountain's strangely pillowed and fissured limestones was to see the reef's surface, faded and discolored by 500 million years above water, but still recognizably zoogenic. The organisms that built such reefs were not the corals of today. They are so long extinct that it is hard to tell what they were. One major reef-building group of porous, conical organisms called archeocyathids was thought to be related to sponges, animals that live by filtering detritus from the water. Now they seem as likely to be related to calcareous algae, which live by making food from sunlight. The creatures that lived in these reefs are equally distant in time, like helioplacoids, spindle-shaped, spirally scaled organisms that perhaps lived in the sea bottom, siphoning food out of the water. Their fossils are common in early Cambrian period western deposits but disappear thereafter.

Archeocyathids and helioplacoids were among the first organisms with hard parts, which appeared in the early Cambrian at the start of the Paleozoic some 550 million years ago. By the mid-Cambrian, such organisms had evolved a startling diversity, and western North America has the most famous example, discovered in 1909 by the same Charles D. Walcott who found the earlier Grand Canyon fossils. Exploring the mountains of eastern British Columbia on summer vacations, Walcott encountered a shale deposit at 8,000 feet on a steep mountainside that eventually, when its layers were split to reveal the fossils therein, yielded some 119 genera and 140 species of Cambrian animals. Because a sudden mudslide had covered the shallow reef where the animals lived, preventing aerobic decay, many of the fossils included soft structures that gave extraordinary insights into the biology of some of the earliest known animals. Stephen Jay Gould, the late twentieth century's paleontological superstar, called the deposit, now known as the Burgess Shale, "the world's most important" fossil fauna.

Despite (or because of) its exquisite preservation, however, the Burgess Shale fauna has caused almost as much confusion as enlightenment about early biodiversity. Walcott considered the Burgess animals ancestral to living phyla such as jellyfish, worms, and arthropods. But some paleontologists who studied the fossils in the 1960s and 1970s found them so strange that they decided many represented extinct phyla, groups as distinctive as living phyla such as mollusks or arthropods. This idea charmed Gould, who used it to construct an evolutionary scenario contravening the historically prevalent notion that life has become more diverse through time. Gould theorized that evolution originally experimented with a great diversity of basic body plans, but that most later died out, leaving the Earth populated by a relatively few lucky ones.

Then, however, new fossil discoveries in places like China and Greenland led some of the paleontologists who had studied the Burgess Shale in the 1970s to change their minds about many of the apparently extinct phyla and link strange fossils back to extant groups. For example, Walcott had classified an oval, scaly little fossil that he named *Wiwaxia* as an annelid worm, and thus a distant relative of the earthworm. Simon Conway Morris, a young English paleontologist who studied *Wiwaxia* in the 1970s, doubted its annelid nature because it lacked wormlike segmentation, and decided that it might have no living relatives. Gould accordingly claimed it as "another Burgess oddball," an extinct phylum. In 1990, however, a researcher Nick Butterfield, found that *Wiwaxia*'s scales had the same microstructure as an earthworm's chitinous bristles, leading Conway Morris to surmise that it was, after all, a very early version of an annelid worm.

Despite the confusion, some Burgess animals have clear affinities with living ones. The second most common animal in the shale, named *Canadapsis*, is a crustacean, a relative of crabs and lobsters, and of the little wood lice common in my garden. Another, a horseshoe crab-like creature named *Sanctacaris*, is considered a chelicerate, a relative of spiders and mites, although it lacked chelicerae, specialized claws that spiders use to manipulate prey. The Burgess fauna also included a few shelled mollusks, although they seem to have been uncommon, unlike the snails in my garden.

Two of the strangest Burgess animals, *Aysheia* and *Hallucigenia*, seem to be onychophorans, an extant group of segmented, wormlike animals, which, however, have multiple legs and may be related to the arthropod group that includes millipedes, centipedes, and insects. *Hallucigenia* is such a bizarre fossil—an apparently headless, segmented cylinder ornamented with bumps, tentacles, and spikes—that Conway Morris at first thought it belonged to an extinct phylum. The spikes are so long that he mistook them for legs and turned the creature upside down. But then new discoveries from China changed his mind. In fact, the Burgess onychophorans were not that much more bizarre than living ones, albeit spikier.

As though to uphold its "world's most important" status, the Burgess Shale also may feature the phylum that includes human beings—chordates. When Walcott named a two-inch-long animal *Pikaia*, he assumed it was a worm because of its flattened and segmented structure. Conway Morris, however, interpreted the segments as bands of muscle called myomeres, a chordate characteristic, and also found that *Pikaia* had a notochord, the stiffened dorsal rod for which chordates were named, and which evolved into the vertebrate spine. *Pikaia* was the same size and general shape as *Amphioxus*, a living chordate of warm coastal sediments, although, unlike *Amphioxus*, it lacked a tail fin.

A mere 20 million years after the Cambrian began, the Burgess Shale seems to have featured all the animal groups that would move onto land, along with many that wouldn't, such as sponges, jellyfish, echinoderms, brachiopods, and trilobites. Similar fossil deposits from a variety of locations, including California and Utah as well as Greenland and China, show that the groups were widely distributed. Given their apparently "explosive" evolution in the sea, a landward movement of these creatures might have been expected to be similarly brisk.

According to evolutionary tradition, however, life didn't colonize the land until the Devonian period, 100 million years after the Burgess Shale. "From the dawn of the Cambrian period, through the Ordovician, to the end of the Silurian . . . the quickening life of the planet remained in the warm primeval oceans," proclaims *The World We Live In,* a classic picture book that awakened my evolutionary imagination fifty years ago. "The dry lands stretched stark and desolate from sea to sea," it goes on, "their drab rocks naked save for a few green films of algae along the shores." In 2000, a PBS *Nova* program about the first tetrapods gave a similar impression of the Ordovician and Silurian periods, showing landscapes of cracked mud and blowing sand. In fact, most land phyla did not leave fossils until later. Land crustacean fossils first appear in Devonian rocks; land mollusks even later, in Carboniferous ones.

There apparently is good reason for the traditional view of life's movement to land as a slow, painful thing. Any walk along a beach, particularly on the west coast, will dramatize this. The actual meeting of land and sea is an abrasive, shifty place of sun, rock, surf, sand, and tides. The animals that live in it—mollusks like mussels and clams, crustaceans like barnacles and amphipods—hide in shells or burrows. They may be incredibly abundant. The rocks below Point Resistance bristle with mussels and barnacles, and the beach seethes with thumbnail-sized amphipods and mole crabs. But there is little landward momentum in their ways of life, which probably have existed since the Cambrian. They are not, in fact, adaptations to land, but to the physical stresses of the littoral zone itself, which reduce predation and competition but require high specialization.

There is a problem with blaming life's slow emergence on littoral zone stresses, however. The traditional view of evolution *doesn't* say that life colonized land across the littoral zone. According to *The World We Live In,* for example, life invaded land not by crossing beaches, but by moving up estuaries and rivers, then into ponds and swamps, before stepping out on *terra firma.* The book contains a two-page panorama by artist James Lewicki of a Devonian estuary teeming with land plants and animals. So the apparent impracticality of a littoral zone route doesn't explain eons of Ordovician and Silurian barrenness.

Moving up rivers and estuaries certainly does sound like an easier route to land than through the surf. But then why did *it* take so long? Moving from seawater into freshwater does pose metabolic obstacles. Freshwater is deficient in the salts that metabolism requires, and animals with permeable tissues would lose body fluids through osmosis without adaptations to prevent this. Indeed, one theory of the origin of hard parts like bones and shells is that they were a way of storing calcium and phosphates in chemically deficient brackish waters. And then, rivers and lakes are less reliable environments than seas, subject to droughts and floods, which could have terminated many promising experiments.

The freshwater route to land may not have been as slow as once thought, however. Recent evidence suggests that some animals began to take it much earlier than the Devonian. Freshwater is a poor preserver of fossils, particularly of very old ones, since its sedimentary deposits are shallow and sparse compared to marine ones. Land is an even poorer fossil preserver. Still, in the 1980s, eastern North American fossil soils from the Ordovician, 440 million years ago, yielded abundant burrows possibly made by millipedes. In the 1990s, British paleontologists found millipede-like tracks in what apparently was the mud of drying Ordovician ponds. Fragments of millipede-like arthropods have turned up from the Silurian period, and other tiny Silurian arthropods left evidence of their existence in the form of fecal pellets containing fungal remains.

Despite the scarcity of this evidence, it is suggestive, since the millipede-like creatures would not have lived in isolation. They would have needed food, plants as well as fungi, and if plants existed, there probably were other plant-eating animals as well, ones that wouldn't fossilize as well as the hard-shelled arthropods.

But were there land plants for Ordovician herbivores to eat? "For perhaps a billion years, the marine plants had drifted virtually unchanged in the primeval seas," says *The World We Live In*. "Then astonishingly, in the space of 50 million years . . . during the Devonian Period . . . they evolved from simple seaweeds into great cone-bearing trees, carpeting the lowlands with ferns and leafy plants, transfiguring the naked hills." A problem with this traditional view, however, is that genetic analysis shows that "seaweeds" such as brown algae are only distantly related to land plants, suggesting that they may not have been their forebears.

Biologists still believe that land plants evolved from algae, but recent opinion has favored not marine ones, but multicellular, freshwater green algae that may have resembled a living group called the Coleochaetales. Presumably, spreading into increasingly dry habitats, these freshwater algae would gradually have evolved features necessary for land life—spores for reproduction out of water, stems for support and transfer of fluids, roots for support and absorption of water and nutrients from soil. Or land plants may even have evolved from land organisms. Moist soils contain single-celled green algae that resemble land plants in chemistry and genetics, and these soil algae probably are as ancient as freshwater algae. A botanist, G. Ledyard Stebbins, thought that "some of these populations of cells evolved directly into flat, tissuelike, multicellular land plants."

Whatever the details, it seems likely that the Coleochaetales-like multicellular algae evolved from a unicellular green algae in either freshwater or soil. Thus, instead of invading from the sea, land plants seem to have been on shore from their green algal beginnings. If this is so, the photosynthetic basis for both land and freshwater ecosystems probably evolved even before the Cambrian.

Indeed, there is no reason to assume that prokaryotes and eukaryotes were confined to the oceans during their first three billion years. By the Cambrian period, they probably pervaded the planet, as they do today, when green algae live inside desert boulders, and prokaryotes inhabit the boiling water of hot springs. At least a rudimentary soil would have covered much of the Precambrian land, barren appearances notwithstanding, and small multicellular organisms—fungi, plants, and animals—could have thronged its moist interstices as they do today. Freshwater habitats also would have teemed with mini-organisms. I once found myriads of tiny jellyfish relatives, hydras, feeding on even tinier crustaceans in a desert rivulet that flows perhaps two weeks of the year. Derek Briggs, who studied the Ordovician animals, observes: "Smaller arthropods may have come ashore at even earlier times, but the smaller the animal, the more difficult it is to find its traces."

Whenever they came ashore, invertebrates definitely were the pioneers and dominated the land long before vertebrates. As herbivores diversified, predators would have proliferated. Devonian sites in Scotland, Germany, and Illinois have yielded fossils of spiders, other arachnids, scorpions, pseudoscorpions, centipedes, and primitive insects as well as millipedes and mites. Other likely early predators were the onychophorans, which, despite their Cambrian marine abundance, live only on land now. Nobody knows how they accomplished this transition, but it suggests great antiquity, although their first probable land fossils are from Carboniferous sites in Illinois and France. I've found onychophorans very like the Burgess fossil *Aysheia* under logs in Costa Rican forest, reddish, velvety creatures that, alarmed at my intrusion, spat a gluelike substance. They use it to catch prey and may have done so since the Ordovician.

Invertebrates still dominate land, particularly insects, which apparently evolved terrestrially from some millipede-like ancestor, since very few are marine. Yet this dominance has a strange adjunct from a coastal viewpoint.

With few exceptions, land invertebrates have failed to reverse evolutionary direction and return to marine life, thus missing opportunities, since there is nothing like the seas for food and living space. Of course, many land invertebrates inhabit coasts, like the kelp flies that can make touring a marine bird or mammal rookery less enjoyable than anticipated. But the ones that actually have returned to salt water are an unimpressive lot. Perhaps this is related to the factors—gravity, gas exchange—that limit their size on land, although sea life might be expected to ameliorate those factors.

Whatever the reasons, invertebrates seem to drop out of the main thrust of coastal evolution after their original land conquests. Not only have land invertebrates shown limited ability to return to the sea, marine ones have seemed little inclined to new land conquest. Despite science fiction's many octopus-like villains, the crafty cephalopods, which appeared in Ordovician oceans, have never even invaded freshwater, much less land. The most octopi do is squirm over rocks from one tide pool to another. Echinoderms and corals have been equally conservative.

Other marine groups—sponges, jellyfish, worms, mollusks—do a lot of slipping in and out of streams. At Limantour Estuary a few miles north of Point Resistance, closed sea anemones stud the sand like pebbles during low tide, a kind of coelenterate pavement. Some mainly marine arthropods may have designs on permanent terrestriality. When I turned over logs on a cloud-forested mountain overlooking Costa Rica's Pacific, I found little brown crabs very like those that scuttle beside Limantour Estuary. Still, invertebrates since the Paleozoic seem to have neglected both sea-to-land transitions and land-to-sea ones, as though having permanent global dominance in animal diversity and abundance is enough. As perhaps it is.

TWO

Amphibious Ambiguities

The Burgess Shale's possible chordate, *Pikaia,* seems a less likely prospect for leaving the water than its invertebrates, having no legs or other traits conducive to land movement. In fact, chordates were among the last phyla to come ashore, perhaps surpassing only mollusks in slowness. According to *The World We Live In,* "the first vertebrate fish originated in lakes and rivers" in the late Ordovician but took until the late Devonian to evolve into "*Ichthyostega,* the first amphibian," a green, torpedo-shaped creature crouched on a log in James Lewicki's panorama. But even our languid phylum may have come out earlier than once thought. Now it seems that "the first vertebrate fish" may have evolved long before the Ordovician, and *Ichthyostega* was not really "the first amphibian."

The ubiquitous Charles D. Walcott discovered Ordovician fish fossils in Colorado in the 1890s, in the Harding Sandstone, a rock formation that occurs in much of the west. The fossils were of ostracoderms, "shell-skinned" fish, a group named a few years earlier by another pioneer American paleontologist, Edward D. Cope. Ostracoderms were very primitive fish. They lacked jaws and had mainly cartilaginous internal skeletons. A shell-like external skeleton covered their heads and much of their bodies, and they ate by filtering sediment from detritus or nibbling with mouthparts formed from their "shell."

Originally, paleontologists thought that the Harding Sandstone was a freshwater deposit, which, along with most vertebrates' dilute blood, convinced them that fish "originated in lakes and rivers." More recently, they decided that the sandstone was marine, suggesting that ostracoderms first appeared in the seas. Since the Harding Sandstone is so widely distributed, some parts of it may have been deposited in estuarine as well as marine water. Still, fully freshwater deposits containing ostracoderms are unknown before the early Devonian, so that group probably did evolve in the sea.

Yet ostracoderms weren't the most primitive fish, although the evidence for this was long one of evolution's mysteries. Two living groups, hagfish and lampreys, actually have more primitive traits. Both groups lack jaws, and they also lack bony skeletons, internal or external. They seem altogether more *Pikaia*-like than ostracoderms. But, while ostracoderms have been extinct since the late Devonian, 375 million years ago, hagfish and lampreys appear later in the fossil record, in the Carboniferous period, 340 million years ago.

Living hagfish and lampreys apparently have survived by evolving specializations that Carboniferous ones lacked, and that exempt them from competition with more recent groups. Hagfish inhabit the seafloor and eat carrion with a protrusible, spiked tongue. They literally tie themselves in knots to get the leverage to tear flesh from carcasses. Adult lampreys parasitize living fish by attaching themselves with sucker-shaped mouths and cutting through the skin with rasplike "teeth," then sucking their body fluids.

Biologists once thought hagfish and lampreys were degenerate descendants of ostracoderms, but later decided that they really are more primitive. A resolution of this apparent paradox finally came to light in the late 1990s, when paleontologists in China found Cambrian fossils from about 530 million years ago, the same time as the Burgess Shale, that resemble hagfish and lampreys. The Cambrian agnathans had rudimentary cartilaginous skeletal features, gills, and fins like living ones, although they were much smaller. So an early "sister group" of hagfish and lampreys evidently did live before ostracoderms but slipped the net of fossil preservation for the following 200 million years.

The Cambrian agnathans turned up in marine sediments, which suggests they originated in the sea. Living hagfish are entirely marine, frequenting deep water and having blood as salty as seawater. Lampreys are more ambiguous, however. All of the world's thirty-nine or so species breed in freshwater. Many are anadromous, the adults feeding in seas or lakes where large fish hosts are abundant, but swimming up streams to breed, each pair building a pebble nest wherein the female deposits hundreds of thousands of eggs before she and her mate die. Young lampreys are tiny larvae called ammocoetes, which live like the primitive chordate *Amphioxus*. After hatching, they burrow into the mud and feed by filtering detritus out of the water. Many lamprey species spend their feeding lives as ammocoetes, because adults live only long enough to breed. Of the ten lamprey species in North America's Pacific drainage, five are nonparasitic and live permanently in streams.

How long lampreys have lived this way is unknown. The oldest fossil lampreys, the 340-million-year-old Carboniferous ones, were in marine deposits, but they were adults and may have bred in streams. Tiny ammocoetes living in shifting bottoms wouldn't fossilize well. So even if they did evolve in the sea, lampreys may have been the first vertebrates to move into streams and lakes, before ostracoderms and other fish. In the late Cambrian and Ordovician, as coastal seawaters filled up with large, competitive invertebrates, soft-bodied vertebrates might have benefited from moving into marginal habitats such as freshwater. Eggs and young would have been less vulnerable. So the dilute nature of lampreys' and most other vertebrates' blood may be a relic of very old movements into estuaries and streams.

The next oldest surviving fish after lampreys are chondrichthyans, including sharks and rays, which appear in the marine fossil record in the late Ordovician. Most species since have lived in the sea, but some have kept moving into freshwater, as though stubbornly, if artlessly, addressing a tempting evolutionary prospect. I've watched from a cliff above Limantour Estuary as countless four-foot leopard sharks swam parallel to the shore and large rays held their own parade a little farther out, waving their "wings" like pennons.

Maybe the parade's purpose was breeding in that protected spot, although leopard sharks, despite a Pacific coast abundance, are mysterious. When Peter Klimley, a shark expert, tried to study a parade he'd seen from a bluff near San Diego, he found it unapproachable, the sharks dispersing at the first hint of his presence in the water. "To this day I do not know the function of this complex and intriguing social pattern," Klimley wrote. He found studying hammerheads in the Gulf of California easier. The Limantour shark and ray parade occurs in August, but both species frequent the estuary year-round. One May, when I stood on the shore as the tide rose, the surface as far as I could see suddenly sprouted ray wingtips and shark tails, a spectacle that lasted a half hour, then stopped.

Although probably first into freshwater, no agnathans or chondrichthyans seem to have crawled out on land. But members of the other main fish group, the bony fishes, have done so repeatedly. The earliest bony fish appeared in the Silurian and quickly attained a bewildering diversity. Some groups are extinct, like the acanthodians, which evolved in the sea in the Silurian but moved into freshwater in the Devonian. They were smallish fish with sharp, bony spines protruding from their fins and other parts of their bodies. Perhaps as an adaptation to preying on them, another extinct group called placoderms had heads armored in bone and sometimes got very big, including an eighteen-foot creature called *Dunkleosteus.*

Placoderms lived in both seas and streams during the Devonian, and one particularly bizarre group called antiarchs may have climbed ashore after a fashion. Antiarch tails were fish-like, but the front halves of their bodies were crustacean-like, their heads encased in boxy jointed armor from which beady eyes protruded. Their pectoral fins, also armored, resembled crab legs connected by a bony joint to the body. These fins, and fossil evidence that one stream-dwelling antiarch named *Bothriolepis* may have had lunglike organs, prompted an American paleontologist, Robert Denison, to speculate that some antiarchs may have pulled out to seek prey on muddy banks.

Surviving bony fish belong to two groups, the ray-fins and the lobe-fins. The vast majority are ray-fins, with fins and tails containing thin, ray supports. Several living ray-fin species spend time ashore. Reconstructions of long-extinct *Bothriolepis* slightly resemble one—the mudskipper, a bug-eyed goby that uses its fins to hop about in Old World tropical estuaries, feeding on amphibious prey. Other ray-fins exploit land life by crawling out bodily, like the "walking catfish," which has escaped aquariums to terrorize Florida suburbs. Others do so by "remote control," like the archerfish, which spits water at terrestrial prey. North America's tropical west coast has the four-eyed fish, *Anableps,* a six-inch guppy relative that has evolved eyes that see in water and air simultaneously, not to watch for prey, but for predators.

Land-crawling ray-fins "breathe" in a variety of ways, by absorbing oxygen through moist skin, mouth, or gill membranes, or by gulping air into lunglike structures called swim bladders, which also serve to regulate buoyancy. (It's unclear which function the structures first evolved to perform.) Apparently, however, no ray-fin fish, living or fossil, has gotten much farther than mudskippers toward a land life. And, considering that they are by far the most diverse and abundant living vertebrates, one may well ask why they should bother.

Most of the few surviving members of the other bony fish group, the lobe-fins, also can "breathe." These are the lungfish, three genera of which inhabit Australia, Africa, and South America. Lungfish don't crawl out on land, although, when their native ponds and streams dry up, the African and South American ones survive by going dormant in mud, using their lungs to supply oxygen. They still need their gills in water, however. The only other extant lobe-fin, the famous "living fossil" coelacanth, *Latimeria,* doesn't use its swim bladder as a lung, since it inhabits deep marine water and can't survive on the surface.

Unknown causes reduced the lobe-fins to their present relict status at about the time when nonavian dinosaurs disappeared. For most of verte-brate evolution, however, they thrived. During the Devonian, lungfish, coelacanths, and other lobe-fins were more diverse and numerous than ray-fins, living everywhere from deep ocean to freshwater and taking many shapes and sizes. One freshwater coelacanth was twelve feet long. Some-thing they had in common, however, was the fleshy lobes that connected their fins to their bodies. The lobes contained bones—a single one con-necting the fin to the body, double ones connecting the lobe to the rest of the fin. This arrangement evidently served them well, although it is hard to say exactly how, since only *Latimeria* and the Australian lungfish have re-tained lobes. The African and South American species have threadlike fins.

However extinct lobe-fins used it, the lobe's skeletal structure has obvi-ous similarities to land vertebrate leg bones, and early evolutionists quickly saw lobe-finned fish as precursors of "the first amphibians." That view pre-vailed until quite recently. *The World We Live In* shows a Devonian lobe-fin named *Eusthenopteron* crawling tiredly out of the swampy estuary over which "first amphibian" *Ichthyostega* languidly presides. With a full set of fins and gills, however, *Eusthenopteron* is still obviously a fish, which seems to call for an explanation as to how and why it might have taken to even part-time life on land.

Evolutionists quickly conceived an explanation, although it did not become widely accepted until 1933, when a Harvard paleontologist, Alfred S. Romer, described it in a popular textbook. Many Devonian fossils are found in reddish sandstones, which geologists considered evidence of desert conditions. Following earlier writers (including the science-fiction novelist, H. G. Wells), Romer theorized that fish such as *Eusthenopteron* had been trapped in drying streams and lakes during massive droughts, and that some had adapted by using their lobe fins to drag themselves from pool to pool, breathing with their auxiliary lungs. Over time, through mutation and natural selection, species would have evolved better ways of moving over-land, such as legs and true lungs.

Romer's idea was so convincing that, in 1938, when Scandinavian pale-ontologists in Greenland discovered *Ichthyostega,* they assumed that it would have been able to stand on all fours. The assumption prevailed until the 1980s, when a group of Cambridge University paleontologists led by Jenny Clack, a museum curator, went to Greenland and found nearly complete skeletons of another Devonian animal, which the Scandinavians had named *Acanthostega* from a few bones they'd found there. Although *Acan-thostega* lived at about the same time as *Ichthyostega,* it was more fishlike, with well-developed gills and a rayed tail. It had four limbs, but Clack and her colleagues were startled to find that instead of having five digits on each—long thought the "primitive" vertebrate digit number—*Acanthostega* had eight, a number unknown in other tetrapods.

When they reconstructed its limbs, moreover, they realized that *Acan-thostega* could not have supported itself on land, or perhaps even crawled out of water. Instead of holding the rest of the skeleton up like legs, its limb bones would have sprawled to the sides like fins. *Acanthostega* evidently had used its limbs and digits not for walking, but for swimming and pushing itself around underwater. When reexamined, *Ichthyostega* also proved to have more than five digits, and Clack thought that although its front limbs probably were strong enough to support it on land, its back ones weren't. Rather than standing on a log, *Ichthyostega* might at best have sprawled on one, like a seal.

In the 1980s, paleontologists also began to question the idea that Devonian desertification had caused the first land tetrapods to crawl ashore. In fact, deserts don't seem likely places for such a development. Modern deserts like California's, at least, haven't caused fish to crawl out of drying pools, although many fish lived in large lakes there during the last glaciation. The lakes disappeared so completely that almost the only fish left are minnows like dace and pupfish, and although some pupfish live in water that dries up annually, like Death Valley's Salt Creek, their response is not to go looking for another creek. Salt Creek is the only water for miles. They die when the creek dries, and their eggs survive in pickleweed, hatching when the water returns.

Paleontologists now think that the red color of Devonian sandstone demonstrates not aridity, but lushness. Tropical forest soils often are red because of leaching and oxidation caused by heavy rainfall. Vascular land plants similar to living club mosses had evolved by the Silurian period, about 400 million years ago. By the Devonian, continents were full of large woody plants, precursors of the even larger club mosses and horsetails

Figure 2. Paleontologists originally assumed that the early tetrapod *Ichthyostega* walked on land, but it probably was mainly aquatic.

whose flammable remains would give the subsequent Carboniferous period its name, "coal-bearing." Neil Shubin and Ted Daeschler, paleontologists who studied Devonian formations in Pennsylvania, replaced Romer's desert paradigm with a swamp one in which legs and feet evolved not as a way of getting out of dwindling water, but of getting around in water full of tree trunks and other vegetation.

Shubin's and Daeschler's work in Pennsylvania complicated things further by turning up a fossil tetrapod that, although slightly older than *Ichthyostega,* might have been more terrestrial. Consisting of skull fragments and a shoulder, the fossil, called *Hynerpeton,* evidently had front legs capable of supporting it and lacked gills, which suggests it might have lived largely on land, although a lack of tail bones made this difficult to confirm. (*Ichthyostega's* finned tail showed it spent time in the water.) *Hynerpeton* suggested that vertebrates might have come ashore well before the late Devonian.

Recently, Shubin, Daeschler, and Farish Jenkins, Jr., a paleontologist at Harvard, complicated things even more by finding a late Devonian lobe-fin that apparently could crawl out on land. In 2006, they announced the discovery on Ellesmere Island in northern Canada of a four-to-nine-foot fossil genus, *Tiktaalik,* which, like *Hynerpeton,* had lived before *Ichthyostega.* Its skeleton suggested that it could prop itself up with its forefins, turn its head in a tetrapod-like way, and possibly breathe with lungs as well as gills. Its skull was flattened like a crocodile's, with eyes on the top instead of the sides. *Tiktaalik's* discoverers surmised that it had lived a semiterrestrial existence, inhabiting streams in what was then a moist, subtropical region, but able to pull out on rocks or mud in a seal-like way. Yet it was definitely a fish, with fins instead of digits (although its bones showed signs of development toward digits). So did legs after all evolve from fins that fish were already using to pull themselves onto land, as A. S. Romer thought—even if not in a desert? Or was *Tiktaalik*'s apparent semiterrestrial life merely a piscine side branch from the more aquatic main stem of tetrapod origins suggested by *Acanthostega*? Only more fossils will tell.

The new evidence still didn't explain *why* tetrapods moved to land. Various possibilities exist. They may have come out to escape aquatic predators, which were formidable in Devonian swamps, including giant, eel-like sharks, or to pursue abundant invertebrate prey themselves. They may have come out to lay their eggs in sheltered places—oxbow pools beside streams, for example—or to sun themselves, as seals and turtles do today. They may even have come out to move from one pool to another, since even the biggest swamps periodically go dry. When I once planned a week of canoeing around the famous Okefenokee in Georgia, there was only enough water for three days.

The new evidence also makes it hard to say just what the "first amphibian" was. Having scales like their lobe-fin relatives, early tetrapods probably resembled them more than modern salamanders. But then, today's "amphibians"—salamanders, frogs, and limbless creatures called caecilians—are specialized creatures that appeared in the fossil record over 100 million years after *Ichtyostega*. They are not transitional between fish and reptiles, but a side branch from the early tetrapod line. Unlike other living tetrapods, they have water-permeable skins, a feature that excludes them from the seawater that covers most of the planet and, in this sense, makes them the least "amphibious" of living tetrapod groups.

A few other Devonian tetrapods are known, but the fossils are not widely distributed. Most have come from the North Atlantic region—Greenland, Scotland, northern Russia. There are a few bones and tracks from Australia, and footprints from South America. No early tetrapods have come from western North America, although fossils show that there was a tropical coast here not too long after *Ichthyostega* began lounging on logs.

The Bear Gulch Limestone in northern Montana is a relic of that coast and rivals the Burgess Shale in exquisite preservation of marine life, in this case, that of a shallow, early Carboniferous bay. Since its discovery in 1968, the limestone has yielded over 113 fish species, including lampreys, chondrichthyans, ray-fins, and lobe-fins. The most abundant was a coelacanth. Most of the species lived at the bay's mouth, which opened on a seaway to the east, from which tides brought nutrients and oxygenated water. The nearshore side supported eel-like, deep-bodied, and filter-feeding species, suggesting that an estuary may have been located there. But tetrapods and other swamp things seem absent. The Bear Gulch Limestone is "unique among Carboniferous fish faunas in the high number of species, the preservation of the fish, and the varied bay habitats," according to Richard Lund and Cecile Popline, ichthyologists who studied it, "and in the complete absence of freshwater macrofaunal elements." Western North America apparently was mountainous and semiarid then, as now. Lacking big swamps, it may have lacked tetrapods, as it lacks big swamps and creatures like alligators today. Or fossils of swamps and their inhabitants simply may not have turned up.

Even if the first tetrapods were confined to big swamps, their evolution seems to have accelerated once they finally got to ground. An early Carboniferous Scottish animal named *Casineria,* discovered in 1999, probably was much more comfortable out of water than Devonian tetrapods. It was six inches long, and thus able to get around the forest floor more nimbly than three-foot *Ichthyostega.* Its toes, reduced to five, could move separately, allowing for better balance and coordination in walking. It may have bred through internal fertilization and laid amniote eggs on land, unlike *Ichthyostega* and its kind, which probably still bred in water. Other fossils believed to be of early Carboniferous reptiles have been found in Scotland and Nova Scotia, along with scorpions, snails, spiders, and nonamniote tetrapods.

"The reptiles, and remains of other organisms, occur in upright stumps of sigillarian trees," wrote two paleontologists. "These animals either fell into rotting tree stumps and were trapped or the stumps were the actual places of habitation. . . . The reptiles were carnivorous and, based on their sizes, dentitions, and co-occurences, the arachnids and insects probably formed their principal food source."

It was such creatures that, once fully adapted to land, would then turn around and go back to the ocean. They seem to have done so with remarkable speed compared to the hundreds of millions of years it took vertebrates to emerge from the water. Fossils of a web-footed, lizardlike swimmer named *Mesosaurus* date from the early Permian period, right after the Carboniferous.

Despite its alacrity, this first return to the sea has been much less a part of traditional evolutionary dramas than the first emergence. Rudolph Zallinger's *Age of Reptiles,* which follows James Lewicki's Paleozoic swamp panorama in *The World We Live In,* is the most famous picture of prehistoric life, viewable on everything from postage stamps to a giant mural in Yale's Peabody Museum. But although it vividly evokes the shift from a sprawling late Paleozoic land tetrapod fauna to an upright Mesozoic one of dinosaurs, pterosaurs, birds, and tiny mammals, Zallinger's picture gives no inkling of how such creatures might have gone to sea.

Of course, fossils of transitional creatures are always much rarer than those of more established organisms. *Mesosaurus* fossils are known only from a few Southern Hemisphere examples, while the *Age of Reptiles* is based on thousands of land and freshwater fossils that accumulated in western North America. Fossils from the late Paleozoic Permian period, when reptiles started returning to the sea, are particularly scarce. North America then was part of a megacontinent, Pangaea, which had formed as crustal movements shoved the planet's landmasses together. Pangaea's formation obliterated shallow, warm inland seas that had existed between continents and that had supported the greatest abundance of marine life. It also obliterated most of the fossils of animals that had lived in those seas.

The Permian doesn't seem to have been a particularly auspicious time for a return to the sea. Not only did steep coasts with cold turbulent waters prevail, its last few million years comprised the deadliest interlude the planet has ever known. Some geologists think an asteroid or comet hit the Earth, raising a dust cloud that cut off sunlight and poisoned the atmosphere. Some think mass volcanism had similiar effects. Fossil sediments suggest that atmospheric oxygen levels dropped drastically, making life much harder. Whatever the reason, an estimated 95 percent of known species, marine and terrestrial, disappeared at the Permian's end, the biggest mass extinction known.

Still, although the traditional onward-and-upward evolutionary drama has downplayed them, Permian marine tetrapods began an evolutionary line that would match more famous land ones in diversity and abundance. Large, specialized marine reptiles had appeared by the early Mesozoic, earlier than dinosaurs. They populated the seas so effectively, moreover, that they evidently kept the dinosaurs out. Although they were the Mesozoic's biggest, fiercest, and probably smartest animals, none of the famous dinosaurs—sauropods, tyrannosaurs, hadrosaurs, ceratopsians—seems to have evolved a marine branch. Only one small, aberrant group went to sea, and they became the subject of one of the earliest and biggest evolutionary controversies arising from North America's west coast.

THREE

Bird Teeth and Reptile Necks

Sitting on the cliff at Point Resistance one warm day, I was startled as pebbles and dust exploded from below me. I thought of earthquakes and landslides, but the cliff wasn't falling. A spiny-scaled little fence lizard was running down it, so fast that I caught only a glimpse before it disappeared into a crevice. A lizard running down a cliff is nothing new, but, fifty feet above the surf, its headlong vertical course was dizzying.

Fence lizards, *Sceloporus occidentalis,* which live throughout the west coast, seem a basic model for terrestrial reptilian life. They sun on rocks, snatch insects, pursue knockabout mating competitions, lay eggs in protected spots. Their habits may resemble those of the earliest reptiles, perhaps even the similarly sized Carboniferous protoreptile, *Casineria.* Edward D. Cope, the pioneer paleontologist who named ostracoderms, likened the first reptiles to "the farm fence lizards of today."

Fence lizards obviously aren't ready to return to the sea, which evokes the mystery as to how early land reptiles managed it. Fence lizard relatives have started to go to sea, however. Little *Sceloporus* is grouped with the herbivorous iguanas, semiaquatic lizards that I've seen diving into rivers on Costa Rica's Pacific coast. They in turn are relatives of Galapagos Island marine iguanas, thought to have originated from Central American ancestors carried to sea during floods. Currents presumably brought them to volcanic islands three hundred miles off the South American coast, where some adapted to feed on coastal rock algae, as marine iguanas do now. Those islands eventually eroded away, but iguanas colonized new ones farther west.

Marine iguanas might be one model for the Permian transition from land reptiles like *Casineria* to water ones like *Mesosaurus*. There are very old, at least partly herbivorous water reptiles—sea turtles, which first appeared in the Triassic period, which followed the Permian. Inferring the past from the present is tricky, however. The odds are against marine iguanas becoming as seaworthy as the penguins and sea lions with which they share the Galapagos. Their population is isolated, not a promising start, and they are not really adapting to the sea but to the littoral zone, where their algae grows. If they ate fish, their chances of getting past the shallows might be better, but given their small size and poikilothermic ("cold-blooded") metabolism, their prospects for invading the deep, cool Pacific around the islands seem dim.

Another living reptile has better prospects, since it eats fish and gets very large. The biggest reptiles I've seen were twenty-foot American crocodiles at the mouth of the Rio Grande de Tarcoles on Costa Rica's Pacific. They resembled bleached logs as they sprawled among the weeds. The Australasian saltwater crocodile gets even bigger and occurs throughout the western Pacific tropics. Both species are mainly coastal, but both occur out to sea. Given millions of years, their descendants might develop specializations such as flippers.

In fact, crocodilians have already gone to sea, many times. Like dinosaurs, they are archosaurs, a group of large reptiles that in the early Triassic already had more efficient gait, musculature, and circulation than lizards and turtles. Early crocodile relatives had evolved a characteristic aquatic form by then. By the Jurassic, at least one group, the metriorhynchids, had webbed feet, a finned tail, and a streamlined shape, suggesting a pelagic way of life. We know that marine crocodiles occurred along North America's western margin in the Jurassic because fragments of skull, vertebrae, and limb bones from another group, the teleosaurs, have turned up in central Oregon. In the subsequent Cretaceous period, some crocodilians, probably marine, grew to fifty feet long.

It's not hard to imagine why crocodilians started moving from rivers to salt water—after fish. But that doesn't explain how some species developed webbed feet and finned tails, nor does it explain why those crocodilians are now extinct, leaving less marine-adapted crocodiles alive. And the limited crocodilian adaptation to the oceans certainly doesn't explain a bewildering variety of more specialized marine reptiles that appeared throughout the Mesozoic, some of which looked and behaved more like fish or dolphins than lizards or crocodiles.

Such questions have loomed large in scientific history. In the late eighteenth century, stone quarriers in the Netherlands found jaws of a giant aquatic saurian, later named *Mosasaurus,* which naturalists first thought had been a crocodile. On examining the jaws, however, Georges Cuvier, the first professional paleontologist, concluded that it had been a giant marine lizard, unlike anything known to be alive. *Mosasaurus* helped convince Cuvier that some species had become extinct, and that the Earth's fauna thus has changed significantly through time. In the next few decades, even stranger, complete, skeletons found in southern English sea cliffs supported Cuvier's then-surprising idea and hinted at a more surprising one.

Cuvier could see no fundamental explanation for what he called life's "revolutions," although he thought catastrophes such as the sudden drying up of seas might cause extinctions. But another naturalist, his Paris museum colleague, Jean Baptiste Lamarck, explained change through the idea of "transmutation" of one life form into another. Lamarck's transmutation resembles the modern concept of evolution but also differs from it. He thought life continually arose through "spontaneous generation" of "animalcules" in soil and water, then transmuted through increasingly complex forms as organisms changed with their environment.

Lamarck could not clearly explain transmutation's cause, however. One explanation he suggested was "inheritance of acquired characteristics," the idea that organisms change by passing on to their offspring traits developed during their own lives. A land animal in an increasingly wet environment, for example, might develop new muscles from swimming and bear offspring somehow possessing such new muscles, eventually transmuting from a legged form to a flippered one. But Lamarck didn't see inheritance of acquired traits as a general cause for transmutation or suggest how traits might be inherited. Cuvier likened Lamarck's ideas to "enchanted palaces of our old romances."

Whatever the causes of life's "revolutions," the new marine fossils were suggestive of transitions. The first whole skeleton found in the English cliffs, by a local carpenter's daughter named Mary Anning and her brother, had so many traits of both fish and reptiles that it acquired the name *Ichthyosaurus,* "fish-lizard." Lacking other income, Anning made a living from finding and selling coastal fossils. She was a remarkable woman, especially considering the dangers and hardships involved in her largely solitary vocation. Her father died from a cliff fall, and lightning struck her as an infant, killing her nurse but transforming her, the story goes, from a dull-witted child into a brilliant one. Her knowledge eventually rivaled that of the gentlemen geologists she supplied, as she sometimes remarked when they condescended to her.

Several years after the *Ichthyosaurus* discovery, Anning found another fossil creature in the cliffs that looked so strange, like a cross between a snake and a turtle, that a geologist, William Conybeare, named it a "near lizard," *Plesiosaurus,* with the idea that it was a marine version of reptiles like snakes or turtles. Conybeare vehemently rejected, however, the "monstrous" notion that a snake or turtle might have transmuted into a plesiosaur. A clergyman as well as a naturalist, he believed that species manifested a series of separate, divinely created forms. But more anomalous creatures kept coming to light. Mary Anning also found specimens of winged creatures, which Cuvier had named pterodactyls, "wing fingers," and, although the French authority deemed them true reptiles, some naturalists thought they seemed transitional with birds or bats.

In the 1820s and 1830s, Richard Owen, England's greatest anatomist, apparently crushed the transmutationists by declaring that none of the strange marine fossils were transitional between reptiles and fish, birds, or mammals—that apparent transitional traits were just superficial similarities. The ichthyosaur's flippers and the pterodactyl's wings were reptilian claws, not transmuted fish fins or incipient bird wings. Owen envisioned ichthyosaurs and plesiosaurs living like crocodiles, crawling ashore to bask and breed, and early illustrations showed this. Owen further denied that one kind of marine reptile could have transmuted into another, whether it be a turtle transmuting into a plesiosaur, or even simply one ichthyosaur species transmuting into another ichthyosaur species. In an 1841 paper, he traced the fossil occurrence of *Ichthyosaurus* from "its first abrupt appearance" to its extinction and saw "no evidence whatever that one species has succeeded or been the result of the transmutation of another species."

In 1859, however, Charles Darwin's *On the Origin of Species* renewed the transmutation debate by proposing what he called "descent with modification" as an explanation for change. Darwin's concept of change, now known as evolution, differed from Lamarck's in two major ways. It saw all life as having descended by heredity from a single primitive organism in the distant past, and it proposed "natural selection" as a general cause for change. Based on how farmers artificially select desired variations to improve domestic stock, Darwin's natural selection theory proposed that organisms with variations providing better adaptation to their environment would be more likely to leave offspring, which would then inherit the variations and be more likely to leave them to *their* offspring, and so on.

A big advantage of Darwin's progressive evolution was that it seemed more testable through the fossil record than Cuvier's catastrophes or Lamarck's transmutation. Primitive organisms presumably would precede better-adapted advanced ones, with transitional forms between them. Darwin cited some fossil evidence for this in the *Origin,* although opponents like Owen dismissed it as inadequate. Within a decade after the book's publication, however, western North America supplied a major arena for testing evolution by yielding rock formations of unprecedented scope. Unlike European formations, they covered vast expanses of semiarid land, rendering the fossil past much more accessible than forested and farmed Europe. The formations encompassed the Mesozoic, and although most Triassic and Jurassic ones were terrestrial, early paleontologists found abundant marine fossils from the Cretaceous, when rising ocean levels inundated the interior for millions of years.

One Cretaceous formation, the Niobrara Chalk, rolls hundreds of miles under the prairie from Kansas to Manitoba, as though the shallow seaway that once linked North America's interior to the Pacific has just drained away, leaving its bottom exposed. Or so it seemed as I walked around the yellow buttes of northwest Kansas's Smoky Hill region one spring day in 1997, picking up apparent marine fossils. A paleontologist later identified some as oyster shells and one, tentatively, as a petrified fece, or coprolite, perhaps that of one of the sea reptiles that demonstrated extinction to Cuvier.

Known since the 1850s, the Cretaceous western sea is so much a part of evolutionary tradition that a Zallinger illustration of its fauna follows his dinosaur panorama in *The World We Live In.* Although the seaway picture lacks the *Age of Reptiles'* scope, showing only a Cretaceous episode instead of marine reptiles' entire 350-million-year story, it is as vivid. Its juxtaposition of an underwater scene of ichthyosaurs, turtles, and fish with a surface one of rearing plesiosaurs and soaring pterosaurs enthralled me at age ten. In the painting's soft aquamarine light, a forty-foot mosasaur, *Tylosaurus,* brandishing a child-sized fish in its jaws seemed more marvelous than scary.

The paleontologists who pioneered the west and discovered many of the creatures in Zallinger's paintings used them to affirm, but also to challenge, Darwin's ideas. Othniel C. Marsh, America's first paleontology professor, was a Darwinian, so when he began western fossil collecting soon after the

Civil War, he was seeking what Owen had denied, transitions between organisms. He immediately found one of history's most important ones.

A major evolutionary controversy at the time was about the origin of birds. Despite their similarities to reptiles, Cuvier had denied any transmutational link, or that birds even had existed in the Mesozoic. The 1861 German discovery of a feathered, winged Jurassic fossil, *Archaeopteryx,* had proved him wrong on the latter point, and *Archaeopteryx* also had a long lizardlike tail and other reptilian traits. Richard Owen still denied it a reptile ancestry, considering it a full-fledged bird. But Thomas Henry Huxley, an anatomist who became Darwin's chief supporter after the *Origin's* publication, found similarities between *Archaeopteryx* and a small dinosaur named *Compsognathus.* He decided that they demonstrated a reptile-to-bird transition.

The lack of one crucial bit of evidence had stymied the controversy. The 1861 *Archaeopteryx* fossil was headless, although there were bits of toothed jaw on the slab that bore the skeleton. Owen thought they were merely fragments of some other animal's jaw. But if they were fragments of *Archaeopteryx's* head, then it evidently had been toothed, an even more convincing reptilian trait than its long tail. Bird teeth became a "missing link" much sought after by aspiring evolutionists.

Marsh worked at prestigious Yale University and had a large inherited income. His position allowed him to lead expeditions of students, scientific assistants, and military escorts throughout the western fossil fields in search of missing links. In 1870 and 1871, his men dug from the Smoky Hill Chalk the bones of a six-foot, flightless marine bird that had inhabited the Cretaceous seaway. (Zallinger's painting shows two paddling beside the rearing mosasaur, *Tylosaurus.*) Although lacking *Archaeopteryx's* lizardlike tail, the bird's skeleton was primitive in many respects, so the implications for the reptile-to-bird controversy were evident. Marsh named it *Hesperornis regalis,* "ruling bird of the west."

Marsh's 1870 and 1871 *Hesperornis* discoveries were headless, like *Archaeopteryx,* but he was a persistent and lucky paleontologist. On an 1872 expedition, he found a "nearly perfect skeleton of this species," which, along with "several other less perfect specimens," showed that *Hesperornis* jaws had indeed possessed the sought-after link. "The maxillary bones are massive," Marsh wrote, "and throughout their lengths runs a deep inferior groove which was thickly set with sharp pointed teeth . . . covered with enamel, and supported on stout fangs. . . . That *Hesperornis* was carnivorous is clearly proven by its teeth, and its food was probably fishes."

Figure 3. *Hesperornis regalis,* the "ruling western bird," had real teeth, unlike all later birds; its discovery in the 1870s provided a "missing link" with reptiles.

Marsh was such a lucky paleontologist, in fact, that he found two kinds of toothed birds in the Kansas Chalk:

> The first species of birds in which teeth were detected was *Ichthyornis dispar,* Marsh, discovered in 1872. Fortunately, the type specimen of this remarkable species was in excellent preservation, and the more important portions of both the skull and skeleton were secured. The remains indicate an aquatic bird, fully adult, and about as large as a pigeon. . . . The jaws and teeth of this species show it to have been carnivorous, and it was probably aquatic. Its powerful wings indicate that it was capable of prolonged flight.

There was some confusion later as to whether some of Marsh's *Ichthyornis* fossils had gotten mixed with reptile ones. But the genus really was a toothed bird like *Hesperornis.* (It too features in Zallinger's painting, soaring overhead of *Tylosaurus.*) Both genera had smaller brains and more reptilian skeletons than living birds, but they were definitely primitive birds.

"The fortunate discovery of these interesting fossils," Marsh crowed in an 1873 paper, "does much to break down the old distinction between Birds and Reptiles, which the *Archaeopteryx* has so materially diminished." Although he found many other important fossils in a long career, none had greater impact. "Nothing so startling has been brought to light since," wrote a younger contemporary, William Berryman Scott. The western discoveries made Marsh America's first celebrity paleontologist, his exploits reported in newspapers like the *Tribune* and the *New York Times.*

Marsh's celebrity culminated when Huxley raved about his toothed birds in an influential 1876 series of pro-Darwin lectures in New York. "Such a discovery," the English anatomist proclaimed, "at once obliges us to modify our definitions of birds and reptiles." Marsh capitalized on the endorsement a year later in a keynote address to the American Association for the Advancement of Science's annual meeting: "The classes of Birds and Reptiles, as now living, are separated by a gulf so profound that a few years since it was cited as the most important break in the animal series, and one which the doctrine could not bridge over. Since then, as Huxley has clearly shown, the gap has been virtually filled by the discovery of bird-like Reptiles and reptilian Birds."

In 1880, Marsh published a lavish monograph on his toothed birds of which he was so proud that he paid for a 250-copy deluxe edition to give to colleagues. On receiving one, Huxley wrote that it "completed the series of transitional forms between birds and reptiles, and removed Mr. Darwin's proposition . . . from the region of hypothesis to that of demonstrable fact." Birds not only had evolved from a group of reptiles but had followed other groups of reptiles in evolving highly specialized marine forms, another transition that Cuvier and Owen had denied. According to Darwin's theory, as the sea inundated the continent, natural selection would have allowed some Jurassic North American birds to adapt to marine life by gradually modifying the gliding wings and perching feet of earlier forms like *Archaeopteryx* into *Hesperornis*'s paddlelike appendages.

Marsh's discoveries were not universally admired. After his first western expedition, the *New York Herald,* the nation's leading daily, carried an article that, although purportedly by a Yale student, expressed sardonic doubts about Marsh's marine fossils: "The vertebrae of the sea snakes in the Yale Museum show the kind of success with which their expeditions were crowned, and we may expect in the scientific journals articles of learned length and thundering sound similar to those which Professor Marsh has already published. . . . In science as well as in fiction, a well-connected tale can be built upon very slight foundations."

And Marsh was not the only paleontologist studying the Kansas Chalk. Although ten years younger, Edward D. Cope, who likened the first reptiles to fence lizards, was hard on his heels. A well-to-do Philadelphia Quaker, Cope was brilliant but volatile, temperamentally opposite to the methodical Marsh. The two had been friendly after they'd met while studying in Europe in 1863, but their relations had deteriorated, largely because of squabbles over marine reptiles. Marsh's shrewdness sometimes made him devious, and after they collected mosasaurs together in New Jersey in 1868, Cope accused him of bribing quarrymen to sequester specimens. Cope's volatility sometimes made him careless, and after he reconstructed the skeleton of a Kansas plesiosaur named *Elasmosaurus platyurus* in 1869, Marsh accused him of putting the head on the tail.

Cope reconstructed *Elasmosaurus* with a short neck and a long tail, but in fact the species had a long neck and a short tail. Zallinger's painting shows it towering over the horizon, neck upraised in a graceful S curve. Seeing Cope's reconstruction in Philadelphia's Academy of Sciences, Marsh "noticed that the articulations of the vertebrae were reversed and suggested gently to him that he had the whole thing wrong end foremost." Cope grudgingly admitted his mistake, blaming it partly on an earlier reconstruction by an Academy of Sciences anatomy lecturer, Joseph Leidy, who, he claimed, had neglected to inform him of the reversed vertebrae. He denied Marsh's contention that "his wounded vanity received a shock from which it has never recovered." Nevertheless, when he made his first trip west in 1871, it was not in a cooperative spirit.

Cope had other reasons to resent the Yale professor. Lacking Marsh's resources, he had to go west alone at his own expense, reaching the Smoky Hill Chalk after the Yale expeditions had combed it for two seasons, and collecting with the few assistants he could afford. He compensated for such disadvantages with an almost hyperthyroid energy. "Marsh has been doing a great deal, I find," Cope wrote his wife, "but he has left more for me." If Marsh had found a pterosaur with an eighteen-foot wingspan (Marsh actually said his pterosaur had a twenty-three-foot wingspan), then Cope found one with a twenty-five-foot wingspan. Other spectacular finds included an eighteen-foot, tarponlike fish he named *Portheus* (now *Xiphactinus*), shown in Zallinger's painting, and an eight-foot turtle, *Protostega*. Cope's determination had an element of bluster, however. He couldn't hope to get Marsh's huge volume of fossils, and he never found anything with the obvious significance of the toothed birds.

But then Cope's ideas of evolutionary significance were not Marsh's. He was religious, which at first inclined him to the creationism some naturalists associated with Cuvier's and Owen's ideas (mistakenly—neither was a creationist). In 1867, he had written that "a great change of temperature" at the Cretaceous period's end had destroyed all animals, followed by "the introduction of new forms of animal life . . . 'the morning of the sixth day' in the Mosaic record of Creation." He soon dropped biblical literalism and had accepted evolution by 1871. But he still denied that a process as random as natural selection was capable of producing the variation that led to major transitions such as the reptile-to-bird one.

In Cope's view, transitional fossils could be taken for granted (he dismissed toothed birds as "simply delightful" in an ironic note to Marsh) but were inadequate to explain the *cause* for transitions. Cope thought he saw a cause, not in Darwin's gradual selection of traits favorable to survival, but in his own version of Lamarck's inheritance of acquired characteristics. He thought such characteristics could be transmitted from parent to offspring during fetal development through a process called "recapitulation," which seemed to him a much better explanation for the way organisms change with their environment than Darwin's more random process.

An idea Cope had about plesiosaurs in 1879 demonstrates his thinking. The various genera had a great diversity of neck lengths, one reason for his mistake with the *Elasmosaurus* reconstruction. "It is known," he wrote,

> that the length of the neck of the Plesiosauroid reptiles of North America diminished as the group approached the period of extinction. Thus the longest necks of this order, those of the species of *Elasmosaurus,* are seen in the Niobrara division of the Cretaceous. In the Pierre formation, we find the shorter necked *Elasmosauri* and the *Cimoliasauri* with still shorter necks. In the latest Cretaceous, the neck is reduced to its most abbreviated proportions in the genus *Uronautes.* The shortening of the neck is thus associated with the shallowing of the water, which, as we know, gradually succeeded the deep-sea period of the Niobrara.

As the sea grew shallower, Cope reasoned, plesiosaurs had less room for neck movement, so each generation would have used its neck muscles differently and passed them on to its offspring, a much faster process than random Darwinian "descent with modification." Physics seemed to support this. In the first edition of his *Origin of Species,* Darwin had estimated the Earth's age at hundreds of millions or even billions of years, allowing plenty of time for natural selection to work. In 1866, however, Lord Kelvin, England's leading physicist, had estimated from the Earth's yearly rate of cooling that it could not be more than 100 million years old, thus allowing only thousands of years for a change from long-necked to short-necked plesiosaurs.

Unfortunately for Cope's example, plesiosaur fossils turned out to show no such neo-Lamarckian development from long-necked to short-necked forms through the late Cretaceous. Long- and short-necked plesiosaurs are scattered through the fossil record from the group's origins to its disappearance. (And dating methods based on post-Kelvin physics have shown that the Cretaceous period alone lasted almost 100 million years.) Cope might have avoided publishing such half-baked ideas if he could have collected as extensively and methodically as Marsh.

But although he could not match Marsh's collecting (an inequity Marsh compounded when, appointed the U.S. Geological Survey's chief paleontologist, he excluded his rival from government funding), Cope bested him in other areas. During his Kansas sojourn, for example, he befriended a publicist named William W. Webb who also disliked institutional Darwinism. Webb soon published an account of a fictive Great Plains exploring expedition, *Buffalo Land,* featuring a pompous "Professor Paleozoic," who leads student trips. Although able to "tell chalk from cheese under a microscope," he mistakes Indian hatchet marks for prehistoric bird tracks. Webb followed his Marsh caricature with two chapters of marine reptile descriptions from "the eminent naturalist Edward D. Cope A.M., who had visited the plains and spent some time in careful exploration."

Although actually less careful than Marsh, Cope was more imaginative. One of his Kansas collecting assistants, Charles Sternberg, recalled that when his boss began to speak of fossil creatures, "so absorbed did he become in his subject that he talked as to himself, looking straight ahead and rarely turning to me, as I stood entranced." The absorption continued in Cope's sleep: "Every animal of which we had found trace during the day played with him at night, tossing him into the air, kicking and trampling upon him." And it showed in his marine reptile descriptions.

"These strange creatures flapped their leathery wings over the waves," Cope wrote of his "twenty-five foot" wingspan *Pterodactylis umbrosus,*

and, often plunging, seized many an unsuspecting fish; or, soaring at a safe distance, viewed the sports and combats of the more powerful saurians at sea. At night-fall, we might imagine them trooping to the shore, and suspending themselves to the cliffs by the claw-bearing fingers of their wing-limbs. . . . Many other huge reptiles and fishes peopled both land and sea . . . far out on the expanses of this ancient ocean might be seen a huge snake-like form which rose above the surface, and stood erect, with tapering throat and arrow-shaped head, or swayed about, describing a circle of twenty feet radius above the water. This was *Elasmosaurus platyurus* Cope, a carnivorous sea reptile.

Cope's visions were not entirely scientific. Pterosaurs had claws like reptiles, not bats, so it is doubtful they "suspended themselves from the cliffs." Although flexible enough to strike at fish, plesiosaur necks perhaps could not "sway about" quite so picturesquely. Cope's imaginings had more popular appeal than Marsh's osteological pronouncements, however, as demonstrated in 1877, when newspapers reported on a living western reptile strongly resembling them. Most were skeptical, albeit playfully fascinated. A *New York Times* editorial on the "Mississippi Monster" joked that it was "a solitary ichthyosaurus which had survived the extinction of the rest of its species, and which has been completely renovated and lavishly decorated for the summer season." The *Tribune* described it as a "pterodactyl-plesiosaur" with a twenty-foot neck, an enormous bill, immense fangs, and a mane of coarse red hair.

The *New York Herald,* mocker of Marsh's sea serpent bones, took the report more seriously. It reprinted a western paper's account of a monster's attack on a barge at a place called "Devil's elbow cut off," during which it supposedly left part of its beak in the vessel. "There are probably some who insist that there had been bending of some elbows beside the elbow of the river . . . ," said a *Herald* editorial, "but we see no reason to be skeptical. . . . There are some queer inhabitants down in the deep waters."

In the next decades, embittered by financial losses, Cope worked imaginatively to undermine Marsh and the U.S. Geological Survey. This culminated in 1890 when the *Herald* published a lengthy attack on both concocted by a hack-writer Cope crony named William Hosea Ballou. In it, Cope accused Marsh of plagiarizing "his alleged work on toothed birds" from assistants, particularly from Samuel Williston, a Kansan who had collected for Marsh and worked at his lab. Williston had begun as a Marsh admirer, searching for toothed bird bones "inch by inch, lying flat on the ground . . . as if they were diamonds." Annoyed when Marsh refused to let him publish on his own, he'd become a Cope informant. "The larger part of the papers published since my connection with him in 1878," the *Herald* quoted Williston as asserting, "have been either the work or the actual language of his assistants."

Marsh vehemently denied the charge, and counterattacked by accusing Cope of lying about his collecting (of saying, for example, that he'd first gone to Kansas in 1870 instead of 1871) and of stealing fossils from various museums, including Yale's. Cope insisted, however, that Marsh's plagiarism was "abundantly proven by the testimony of the men who wrote the books," and would be "further substantiated by them and by others in the *Herald.*"

Further substantiation proved scanty. The *Herald* dropped the affair, disappointed with its failure to scandalize the public, and the scientific establishment deplored it. The Philadelphia Academy's Joseph Leidy, a meticulous naturalist who had abandoned western paleontology because of his younger rivals' unseemly squabbles, told the press that Cope's behavior had "caused the deepest regret among the scientific men of the country." Still, it returned to haunt Marsh two years later. The Geological Survey and its director, John Wesley Powell, had powerful enemies in Congress, and they used the *Herald* slurs in a campaign to cut survey appropriations, citing Marsh's deluxe toothed bird monograph as proof of government extravagance, although he had financed it himself. Marsh's survey job was among the cuts, and his lavish collecting ended.

The stress-ridden Cope met an early death in 1897, but Ballou continued to torment Marsh with magazine articles attributing the major Mesozoic reptile discoveries to his rival. "Cope, perhaps, defined the greatest number of species," an 1897 piece declared of marine reptiles, allowing Marsh only the definition of a few skeletal details. According to Ballou, Marsh called on the editor of *Popular Science Monthly* to demand that he not publish an 1898 marine reptile piece of Ballou's, "as I had grossly misrepresented his work," but the editor replied that "the monthly had the habit of printing what it pleased and preferred my work to his and would print it." Marsh "went away in a huff" and succumbed to a stroke in 1899.

Cope's plagiarism charges remained unsubstantiated. His colleagues regarded his opposition to Darwinism more sympathetically than his rivalry with Marsh, however. After Darwin's death, natural selection fell into disfavor, because, as Cope had said, it did not explain the variation on which selection would have to operate. Darwin's theory returned to vogue after genetics began to explain variation and physics changed Lord Kelvin's 100-million-year-old Earth back to Darwin's much older one. But that took until the 1920s, and, of course, debate about Darwinism continues.

Still, evolutionary transition has seldom been displayed more vividly than it was on the dessicated remnants of the western interior sea over a century ago. After receiving a copy of the deluxe toothed bird monograph in 1881, Darwin himself wrote Marsh: "Your work on these old birds and the many fossil animals of N. America has afforded the best support to the theory of evolution that has appeared within the last twenty years." Even the die-hard anti-Darwinian Richard Owen called the monograph "the best contribution to natural history since Cuvier."

FOUR

Tail Tales

Despite Darwinism's hiatus, western marine fossils kept suggesting evolutionary transitions. Bones found in California disproved Richard Owen's contention that the swordfishlike ichthyosaurs had appeared in the Jurassic and changed little during their existence. The bones' scientific discoverer, John C. Merriam, resembled Charles D. Walcott in belonging to the paleontological generation following Cope and Marsh, and in looking for marine fossils in some of the west's steepest, driest environments.

Merriam got his PhD studying marine reptiles and taught at UC Berkeley between 1894 and 1920, then a less well-funded position than the USGS. In 1900, however, an energetic young heiress named Annie Alexander attended his lectures and became so enthralled with paleontology that she began financing expeditions to places that might have been inaccessible otherwise. Alexander eventually financed UC Berkeley's Museum of Paleontology, after Merriam, whose temperament resembled his imperious predecessors', abruptly departed for a loftier position at Washington's Carnegie Institute.

One of the west's steeper places is the Klamath Mountains, a New Hampshire-sized knot of peaks and gorges in northwest California and southwest Oregon. In 1893, James P. Smith, a geologist looking for ammonites, shelled cephalopods, found apparent reptile bones in late Triassic limestone at the Klamaths' eastern edge. Crustal stresses have tortuously folded and upended the 205-million-year-old strata, called the Hosselkus Limestone, so Smith didn't find an entire skeleton, only vertebrae, ribs, and limb bones. He thought they were from a plesiosaur, but Merriam decided, after studying the bones, that they were part of an ichthyosaur, which he named *Shastasaurus pacificus.* On condition that she could go along, Alexander financed a number of trips to the area's limestone cliffs in the next decade, among the earliest scientific expeditions in which women actively participated.

Alexander and her female companions (propriety forbade an unmarried woman to camp alone with men) found many fossils and seem to have been more competent explorers than some of the professionals. An assistant professor negligently set the camp on fire, while Merriam once had to dissuade his fossil preparator from attacking a large bear and her cubs with a .38 caliber pistol. Despite such excitements and others, including big rattlesnakes, numerous mountain lions, and unruly packhorses, the expeditions collected some two hundred Triassic reptile fossils in the Klamaths, providing a good idea of *Shastasaurus*'s appearance.

From an evolutionary viewpoint, the most significant thing about its appearance was the tail. The English Jurassic ichthyosaurs had a sharp downward bend at their vertebral column's end, and others in German shale beds suggested the reason for it. The shale sometimes preserved body outlines, showing that Jurassic ichthyosaurs were more fishlike than Richard Owen had thought, with dorsal fins and bilobed tail fins of which the downturned vertebrae formed the main support. The earlier *Shastasaurus,* on the other hand, had only a slight bend in its tail, suggesting that its tail fin had been more rudimentary than Jurassic ichthyosaurs', more like a lizard's than a shark's. *Shastasaurus* also may have lacked Jurassic ichthyosaurs' dorsal fins, although no fossils show its body outline. Other late Triassic genera that Merriam examined from the Klamaths and sites in Nevada also had primitive traits. So ichthyosaurs had undergone significant "descent with modification" during their 150 million years of existence, evolving fishlike tails and other traits that evidently increased their marine fitness.

The Klamath Mountain skeletons did not reveal ichthyosaur origins, which remain unknown. Although presumably evolved from land reptiles, even the earliest known ichthyosaurs, slender, yard-long, early Triassic creatures named *Utatsusaurus* and *Chaohusaurus* and found on Japan's northeast coast in the 1980s, probably never came ashore to bask as Owen imagined. Despite rudimentary tail fins, they evidently were fully marine,

Figure 4. *Shastasaurus,* a Triassic western ichthyosaur, was fully aquatic, but it looked less fishlike than later genera. It may have lacked a dorsal fin.

sharing with later genera a striking pelagic trait. As early as 1846, adult ichthyosaur skeletons began to turn up with small, probably fetal, ichthyosaurs inside. Even Owen allowed that, instead of laying eggs ashore like sea turtles, they might have incubated internally and given "live" birth, as do many snakes and lizards.

Other western ichthyosaurs evidenced an inability to come ashore. In the 1890s, miners in the central Nevada town of Berlin dug up fossil vertebrae so large they allegedly used them as dinner plates. When a geologist scouted the area in 1928, he discerned that the vertebrae belonged to late Triassic ichthyosaurs, which, when excavated in the 1950s by UC Berkeley paleontologists, proved to be a fifty-foot animal that they named *Shonisaurus* after the local Shoshone Mountains. Such a creature could no more have dragged itself up a beach against the force of gravity than could a humpback whale. Berlin is a ghost town now, but at the present Berlin-Ichthyosaur State Park, eroded, scattered remains of around forty *Shonisaurus* are preserved *in situ,* giving a sense of their mass, especially in their present high and dry location.

Like *Shastasaurus, Shonisaurus* may have lacked a dorsal fin and had a relatively rudimentary tail fin. Its head, consisting largely of narrow beaklike jaws, must have looked peculiar in front of its huge body, the ribs of which were taller than a grown man. The six-foot-long fins were also oddly narrow, and the teeth were surprisingly small and may have grown only in the front of the jaws, so that it is hard to imagine how *Shonisaurus* got enough to eat. The skeletons are aligned in a row, which at first suggested they might have beached themselves as whales sometimes do. But central Nevada was deep ocean in the late Triassic, so their apparent mass death remains unexplained.

Although John Merriam and Annie Alexander didn't find *Shonisaurus,* going to see it even today shows what a windfall Alexander's funding must have been. It's still a long dirt detour from U.S. 50, itself called "the loneliest road in America," and I got a shredded tire on the way. But that is nothing compared to a seventy-five-foot *Shonisaurus* skeleton discovered in the 1990s at a site beside British Columbia's Sikanni Chief River, so far from any road that the heavy equipment to excavate it had to be brought in by helicopter. The stone block containing its skull weighed four tons. It is the largest marine reptile known.

Something that Merriam and Alexander did find seems an ironic footnote on the ichthyosaurs' Triassic appearance as full-blown pelagic animals. The Klamath Triassic limestone originated as part of a complex of islands and reefs off the west coast of what was then the supercontinent of Pangaea. Along with ichthyosaurs, the Berkeley expeditions found bones of reptiles that apparently had lived on the islands' shores and reefs. One genus, *Thalattosaurus,* grew six feet long, with webbed claws, a finlike tail, a strangely hooked snout, and knobbed teeth that perhaps were adapted to eating ammonites, common fossils in the limestone. Another, *Nectosaurus,* had less robust teeth and perhaps was a fish eater. Well-developed legs and claws suggest that the reptiles were as nimble climbing on steep rocks as they were swimming in warm seas. So as ichthyosaurs swam like dolphins and whales in the late Triassic ocean around the islands, thalattosaurs, which perhaps resembled the ichthyosaurs' mysterious Permian forebears, were just starting a marine career nearby.

Something went wrong for thalattosaurs, however, and instead of becoming a new pelagic reptile group, they vanished after the Triassic. Such evolutionary twists are common, as if a creator, making fossils to test man's faith, as some biblical fundamentalists believe, had lightened the task with puzzles and whimsy. Ichthyosaurs, for instance, might have been expected to prosper as they became more fishlike, increasing in diversity and branching into new habitats. North America's west coast has produced more Triassic genera than later Jurassic ones, however, and the group generally seems to have become less diverse through the Mesozoic, not more. By the late Cretaceous, only a large swordfishlike genus, *Platypterygius,* survived, and despite its advanced marine adaptations, it died out some 28 million years before the Mesozoic's end. Some paleontologists theorize that competition from increasingly formidable fish such as swordfish and tuna may have contributed to its disappeance, but nobody is sure.

Ichthyosaurs weren't the only west coast creatures to vanish mysteriously millions of years before the dinosaurs. Marsh's toothed bird genus, *Hesperornis,* clearly was well adapted to sea life, swimming nimbly with broadly lobed feet, snagging fish with its toothed bill. It was so abundant

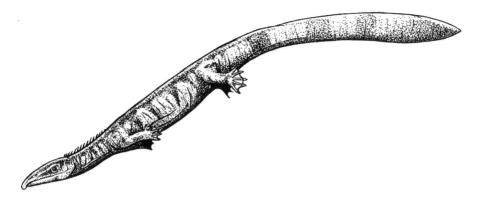

Figure 5. Thalattosaurs, "ocean reptiles," inhabited western shorelines and reefs in the Triassic period but died out before they evolved a fully aquatic form.

that it left numerous fossils as far north as Canada's Northwest Territories. But *Hesperornis* seems to have left no fossils after about 70 million years ago, 5 million years before the dinosaurs' end, although a few Montana fossils suggest that a freshwater relative named *Potamornis,* "river bird," may have survived into the immediately postdinosaur age.

Pterosaurs also languished in the late Mesozoic, although a few may have lasted to the end. The twenty-five-foot-wingspan genus Cope and Marsh squabbled over, now called *Pteranodon,* apparently did not survive that long, but another genus named *Quetzalcoatlus* may have. Discovered in Texas in 1971, it had a thirty-six-foot wingspan. That such monsters out-lived most of the toothed birds seems another evolutionary whimsy, especially as it remains unclear how pterosaurs managed to live at all. *Pteranodon* skeletons have been found with fish fossils, implying they ate fish. Given their size, fragility, and toothless "beaks," it's hard to imagine them eating anything larger. But nobody knows how they caught fish. They may have snatched them while flying over the surface, like fishing bats, but there is no evidence to prove this. Their leathery wings probably didn't let them "dive-bomb" like pelicans or "fly" underwater like murres and cormorants. They may have speared fish while floating on the surface, but it is not clear that they could float.

Most known *Pteranodon* fossils are from interior seaway deposits, showing that they at least flew over water. Fragmentary pterosaur fossils, probably of *Pteranodon* and a *Quetzalcoatlus*-like species, have turned up in California and Oregon, proving that they frequented the eastern Pacific as well. Other *Quetzalcoatlus* fossils are associated with deposits hundreds of miles inland, so *Quetzalcoatlus* evidently frequented land or freshwater as well as the sea. If so, it may have fed heron-fashion, stalking about to spear fish in shallows or small reptiles and mammals in marshes.

At least some pterosaurs probably nested in rookeries. Kevin Padian, a paleontologist at UC Berkeley, found what may have been a pterosaur rookery in the South American desert, a large deposit of juvenile bones accumulated by a Mesozoic flood. That rookery evidently was inland when pterosaurs used it, but, although there is no fossil evidence of this, they probably bred on coasts and offshore islands too, perhaps with toothed birds, since islands and other protected sites are always in short supply.

Despite theories, nobody knows why ichthyosaurs, toothed birds, and pterosaurs disappeared before other, equally ancient marine tetrapods. Turtles, one of the oldest tetrapod lineages, had evolved marine forms by the mid-Triassic. Their need to lay eggs ashore might seem a liability compared to pelagic live-bearing, but it clearly wasn't. Plesiosaurs also had appeared in the Triassic, having perhaps evolved from a group of amphibious reptiles called nothosaurs. They probably bore live young in the water, although fossil evidence is inconclusive. Fossil nothosaur embryos not enclosed in eggs have been found in China and Europe, but they are not enclosed in adult nothosaur skeletons. The size of genera like *Kronosaurus*, a forty-foot-long, short-necked plesiosaur in Zallinger's painting, suggests that they were fully pelagic, since such giants would have been as incapable of coming ashore as the fifty-foot ichthyosaur, *Shonisaurus.*

As the last, swordfishlike Cretaceous ichthyosaurs dwindled, ironically, creatures like the more saurian-looking Triassic ichthyosaurs were thriving along the west coast. Restorations of a Cretaceous mosasaur named *Plotosaurus* from the Moreno Formation in the arid hills west of California's San Joaquin Valley startlingly resemble Merriam's Triassic ichthyosaur, *Shastasaurus.* Both had small, short-beaked heads and rudimentary tail fins. The resemblance was a result of "convergent" evolution rather than divine whimsy. Organisms occupying similar ecological niches, although distantly related, often evolve similar adaptive features. The toothed bird, *Hesperornis,* for example, resembled mosasaurs in having a jointed lower jaw that helped it to swallow fish whole.

Mosasaurs are unusual among marine reptiles in that their origins are known. Their skeletons resemble those of varanid lizards, the group including today's monitors, which are good swimmers. I once saw a monitor dive with startling dispatch from a tree into a southeast Asian river. Mosasaurs

Figure 6. *Plotosaurus,* a giant marine lizard related to living monitor lizards, inhabited the west coast just before the late Cretaceous mass extinction.

probably evolved from amphibious varanids in the mid-Cretaceous, perhaps sharing an ancestor with a monitor-like group called aigialosaurs. But life in the Cretaceous ocean evidently was enough like life in the Triassic one that *Plotosaurus* evolved to look like *Shastasaurus.*

Mosasaurs probably gave live birth at sea too. In 1996, paleontologists tentatively announced the discovery of a large skeleton associated with two small ones in the Pierre Shale of South Dakota. According to their abstract, the small mosasaurs had "prenatal" characteristics and showed no sign of having been digested in the big one's stomach. In 2001, two paleontologists described embryos in a European aigialosaur skeleton, suggesting that mosasaur ancestors had evolved live birth even before marine traits like flippers.

Evolutionary convergence is a powerful force that affects habitats as well as inhabitants. A Cretaceous scene on the western North American coast might have looked not unlike Point Resistance today, with forested ridges running down to stony headlands. There would have been plenty of tectonic activity, earthquakes and rising land, as a Pacific oceanic plate subducted under a continental one. Cliffs perhaps overlooked the shore, and a time traveler on one might have seen pterosaurs or toothed birds resting on offshore rocks. Fossil sites like the hills west of the San Joaquin Valley show that marine reptiles were abundant on the late Cretaceous coast. Along with two species of the mosasaur genus *Plotosaurus,* the Moreno Formation has yielded bones of another, larger mosasaur, two kinds of plesiosaurs, and a marine turtle. So whalelike shapes would have passed offshore, mosasaurs migrating toward breeding or feeding grounds. A plesiosaur's or turtle's head in the shallows might have resembled a seal's.

There may have been other similarities with Point Resistance. Zallinger's seaway painting shows only toothed birds and pterosaurs, but DNA studies and fossils suggest that some modern bird groups lived in the late Cretaceous. The fossils are scarce. Of 150,000 late Cretaceous to early Tertiary fossil specimens collected in Montana and Alberta by the UC Berkeley Museum of Paleontology, only a few dozen are of birds. And most are too incomplete for clear identification. In 2003, California's Mesozoic record of possible still-living avian groups consisted of a limb bone from a pigeon-sized bird, otherwise unknown, although possibly marine, since it was in a coastal deposit.

Still, such fossils suggest that several still-extant bird groups might have existed in the late Cretaceous, and these groups, whimsically, may include some that frequent Point Resistance today. In 2005, a group of scientists judged an Antarctic Cretaceous fossil named *Vegavis* to have been an anseriform, a relative of the scoter ducks that ride the surf. Other Cretaceous fossil birds may have been gaviiforms, related to the loons that also ride the surf. A fossil genus called *Cimolopteryx,* found in Wyoming, may have been a charadriiform, and thus related not only to Point Resistance's murres, but to its guillemots, gulls, terns, oystercatchers, and sandpipers.

The fossils do not prove that birds resembling today's scoters, loons, or murres existed in the Cretaceous. Indeed, their scarcity suggests that the modern bird groups were rare then. But distant relatives of ducks, loons, and murres may possibly have lived with *Hesperornis* and *Ichthyornis,* and an interesting possibility arises if so. Marine rookeries like the Point Resistance murre rock are stressful. Competition for space is intense; exposure to the elements is harsh; predators pick off eggs and chicks. Cretaceous rookeries would have been similar. If modern bird groups started climbing aboard, the rookeries would have become particularly stressful, suggesting one Darwinian explanation for dwindling toothed birds and pterosaurs— that modern birds crowded them out.

Such evolutionary struggles wouldn't be any more evident than today's to a time traveler. Notions of prehistory as a time of teeming, unrestrained savagery are imaginary, since organisms lived under the same environmental constraints as now. A venerable somnolence like the one at Point Resistance would have pervaded its Cretaceous counterpart, a greater one, since the marine reptiles' world had by then lasted almost 200 million years. Much was about to change, however, as the Cretaceous neared its end.

Catastrophes seem to have converged on North America's west coast then. The Earth underwent one of its more destructive episodes of tectonic activity, with Indian volcanoes spewing forth enough ash and lava, if equally distributed, to cover the planet. Volcanoes were active around North America's inland sea, as vapor-wreathed cones in Zallinger's painting suggest. Tectonic activity warped the continent upward, draining interior seas back into ocean basins, and cooling climate. Clouds of volcanic ash and toxic gases polluted the atmosphere. Earthquakes caused epic tsunamis and floods. Eruptions ignited vast forest fires, with their own air-polluting effects. Pollution returned to the surface in acid rain and other forms.

Bad times can always get worse and evidently did 65 million years ago. A crater 100 miles wide beneath Yucatan and the Gulf of Mexico shows that an asteroid or comet roughly the size of Mount Everest landed there, with effects analogous to multiple thermonuclear blasts. The impact could have caused an earthquake, tsunami, and ash cloud temporarily dwarfing mass volcanism. The cloud might have obscured the atmosphere long enough to cause global darkness, and its toxic materials might have temporarily poisoned land and water.

Whatever happened, the fossil record shows it was lethal. A layer of clay virtually devoid of organic remains overlies late Cretaceous strata in many places. In western North America, relatively close to the Yucatan impact, various strata suggest dramatic events. California's Moreno Formation contains tektites, small glass spheres apparently blasted out by the impact. At a place called Crowley's Ridge in Missouri, a terminal Cretaceous deposit contains a mosasaur tooth and vertebra along with tektites and jumbled rocks. It may be a residue of the impact's tsunami, a wave as tall as a thirty-three-story building. Above that stratum is a "virtually unfossiliferous" limestone layer, characterized by paleontologists as a "dead zone."

Many organisms, marine and terrestrial, disappear above "dead zone" strata. These include most of the species of microbes, such as cocoliths and foraminifera, that had lived on the ocean surface, and many larger organisms, including ammonites and lobe-finned fishes, except the surviving coelacanth. Some mollusk, crustacean, and fish groups that did reappear had dwindled sharply in diversity and abundance. Since surface plankton and the fish and invertebrates that fed on it were basic to the food chains that supported mosasaurs, plesiosaurs, and pterosaurs, their decimation may help explain why the giants disappeared.

On the other hand, except for the Missouri tooth and vertebra, the most recent known fossils of mosasaurs, pterosaurs, and plesiosaurs are many thousands of years older than the dead zones, so there is very little direct evidence that an impact killed them. And most marine invertebrate and fish groups reappeared above the dead zone strata, as did crocodilians, turtles, and birds. Their survival is as little understood as the others' extinction. Crocodilians, turtles, and birds lay eggs on land, but so did pterosaurs.

Many species frequent freshwater as well as salt water, but so, possibly, did some mosasaurs and plesiosaurs. More evidence is needed to explain the extinctions, and it is hard to imagine what it will be.

Despite the mysteries, a pattern of tetrapod land-to-sea evolution seems to emerge from the Mesozoic era. Groups that went to sea tended to do so early in their evolution. When dinosaurs and early mammals were just developing characteristic land adaptations in the late Triassic, turtles, ichthyosaurs, and plesiosaurs had already evolved fully pelagic marine forms. Even the late-coming mosasaurs seem to have gone to sea early in the development of their own lizard group. *Hesperornis, Ichthyornis,* and their relatives were among the earlier birds, and ancient relatives of ducks, loons, and murres are among the earliest known modern birds.

After groups went to sea, they evolved differently in some ways from terrestrial ones. While dinosaurs and mammals underwent major diversifications of form and function from the Triassic through the Cretaceous, ichthyosaurs, plesiosaurs, and mosasaurs kept closer to basic patterns. Pterosaurs evolved from long-tailed, short-winged forms in the Triassic and Jurassic to short-tailed, long-winged forms like *Pteranodon* in the Cretaceous, but their anatomy otherwise didn't change radically. Marine birds lost teeth and got bigger brains but maintained loonlike and ternlike forms. Marine turtles varied less in size and shape than terrestrial ones. Diversity seems to have peaked toward the middle of each marine group's existence, then dwindled gradually thereafter.

There were exceptions. Plesiosaur diversity was still fairly high when the Cretaceous extinction occurred. Mosasaur diversity was increasing. The California genus, *Plotosaurus,* is the most advanced form known and probably swam better than earlier mosasaurs, judging from its unusually rigid spine. Turtle and bird diversity may have been increasing in the late Mesozoic, but it is hard to tell from available fossils.

Whether leading to extinction or survival, anyway, the land-to-sea transition seems to have been a one-way street for Mesozoic tetrapods. Marine pterosaurs may have flown to nesting sites inland and fed partly on land animals, as some marine birds do today, but such adaptations don't necessarily lead back toward a land life. Ichthyosaurs, plesiosaurs, and mosasaurs may have swum into lakes and rivers, but they don't seem to have evolved back into strictly freshwater forms, much less into amphibious ones. If they couldn't crawl out on a beach, the chances of doing so in a swamp or forest were slight. Still, there may have been an exception to this pattern too, one involving mosasaur relatives.

FIVE

Cope's Elusive Ophidians

Walking along the coast trail near Point Resistance one foggy May afternoon, I noticed what looked like a piece of brown bungee cord in the dust. When picked up, it flexed, curling around my little finger and revealing a yellow underside. It was a Pacific rubber boa, a small western version of the neotropics' big boa constrictors. Before releasing it, I examined the end that was constricting my finger and saw two spurlike appendages, vestigial hind limbs that males use, not for locomotion, but to grasp females during mating.

As their vestigial limbs suggest, boas and their Old World relatives, pythons, make up one of the oldest groups of living snakes. They were particularly numerous in the Paleocene epoch, which followed the Cretaceous mass extinction, as shown in Rudolph Zallinger's *Age of Mammals,* a panorama of the past 65 million years that follows his marine reptile painting in *The World We Live In.* The most arresting image on its far left, which shows the Paleocene, is a great boalike snake dangling from a tree. It seems to mark the transition between the reptile and mammal ages, and snake evolution does link the two in that snakes are the youngest, most vigorously diversifying reptile group. It is strange to think of a world without snakes, but they did not appear until the Cretaceous.

There has long been a majority view among herpetologists as to how snakes evolved. They are classed in the same order as lizards, Squamata, and today numerous lizard species live by burrowing underground and have reduced or atrophied limbs. Some of the most primitive snakes are burrowers, so a reasonable explanation of their limblessness is that they evolved from burrowing lizards, then moved back to the surface and, finding a niche as nimbly slithering predators on small animals, prospered. Even the harmless-looking rubber boa preys on lizards and mice by constriction.

There has also been a minority view on snake origins, and the old troublemaker, Edward Cope, started it. The lizard group to which snakes are most closely related is the varanoids, the mosasaurs' close relatives, and Cope decided that mosasaurs were essentially snakes with flippers. "These animals," he proclaimed through his mouthpiece, William H. Ballou, in an 1897 *Century Magazine* article, "suggest the fabled sea serpent . . . in fact, they possessed eight technical characteristics of serpents of the same period. . . . The teeth, without fangs, are those of serpents, and they differ from the teeth of any of the lizards. . . . Science must regard the mosasaurs and their allies as a race of gigantic, marine serpent-like reptiles, with powers of swimming and running like modern snakes." In the 1898 *Popular Science Monthly* article on marine reptiles that so enraged Marsh, Ballou wrote that mosasaur skulls showed they had hissed and protruded forked tongues, while bits of mosasaur "scales and skin, found perfectly preserved" also were snakelike. "Professor Marsh formerly thought, and it has been taught in classrooms, that the bodies of mosasaurs had bony scales," he jeered. "They had skins and were scaled throughout like modern lizards and snakes."

Childhood enthusiasms influenced Cope's idea. He showed an early fascination with marine creatures, as when, at age seven, he kept an illustrated journal of a voyage from Philadelphia to Boston. "We saw some Bonetas swimming along-side the vessel; they are long slim fish & they twist about like eels," he wrote, with precocious exactitude. He described orcas as vividly, using their original common name, more appropriate than "killer whale," since orcas do kill other whales: "Today some whale killers came near. They are large black fish and they blow water out of their heads. Some of them have white spots on their sides. One came close along side the vessel. The captain ran and got a harpoon to catch one, but it was too late they had all swam away." Young Edward also drew a precisely observed orca spout and dorsal fin.

Such sights must have recalled an exciting discovery of the previous year. "I have been at the Museum," six-year-old Edward wrote to his grandmother in 1846, "and I saw Mammoth and Hydrarchas, does thee know what that is? it is a great skeleton of a serpent. It was so long that it had to be put through three rooms." The imaginative child was not exaggerating. The "Hydrarchas" skeleton, exhibited at Philadelphia's Peale Museum by a showman named Albert Koch, was a 114-foot chain of keglike vertebrae ending in a 6-foot skull with yam-sized teeth. Koch had found the bones in Alabama the year before and had displayed them in New York, impressing newspaper reporters if not an anatomist named Jeffries Wyman, who observed that the vertebrae were remarkably diverse in size and condition to have come from a single animal. In fact, Koch had built the skeleton, which he named *Hydrarchus,* "water ruler," from several wagonloads of scattered bones. It eventually proved to contain the remains of five individuals belonging to more than one species.

Koch showed some effrontery in taking *Hydrarchus* to Philadelphia, since the real genus from which he'd concocted it had long been known at the American Philosophical Society there. In the 1830s, the Society had received similar bones from marine deposits formed when a later remnant of the Cretaceous interior seaway still covered much of Louisiana and Alabama. William Harlan, an anatomist, had examined them, and thought the vertebrae and jaws reptilian, indeed suggesting a "sea serpent." The skull puzzled him because its cusped back teeth seemed more mammalian than reptilian, but he had concluded that it was a thirty-foot marine reptile and had named it *Basilosaurus,* "ruler lizard."

Harlan's paper on the bones had incited another round in the fight between Cuvierians and transmutationists that earlier involved ichthyosaurs. Transmutationists had hailed *Basilosaurus* as a link between reptiles and mammals, but after Harlan brought the fossil to England in 1839, Richard Owen had again crushed them by demonstrating that its supposed reptilian traits were superficial and its teeth definitely mammalian. Noting that its hollow jaws resembled a living sperm whale's, Owen identified *Basilosaurus* as an early cetacean and changed its name to *Zeuglodon,* "yoke-tooth," in recognition of its double-rooted molars. (That violated the rules of priority, however, and the name has returned to the original, Latin for "ruler lizard.")

The grown-up Edward Cope, of course, knew that the "Hydrarchas" of his childhood wonderment was a hoax. He collected and described whale fossils, including *Basilosaurus-Zeuglodon.* That knowledge apparently didn't expunge an imprinted enthusiasm for seeking links between serpents and sea creatures. The enthusiasm budded in the midst of early feuding with Marsh. "We may now look upon the mosasaurs and their allies as a race of gigantic, marine, serpent-like reptiles, with powers of swimming and running, like the modern Ophidia," he proclaimed in 1869, during their squabbles over New Jersey fossils. "Thus in the mosasaurids, we almost realize the fictions of snake-like dragons and sea serpents, which men have ever been prone to indulge. On account of this ophidian part of their affinities, I have called this order the Pythonomorpha."

The very next year, Marsh came up with a giant Paleocene snake, *Dinophis grandis*, "great terrible serpent," from New Jersey. "[I]n this country," he announced, "two species only, one founded on a single vertebra, have been described hitherto," adding that his own specimen, evidently much more than a single vertebra, indicated "an animal not less than thirty feet in length; probably a sea serpent allied to the Boas of the present era." Marsh closed his article with a typical circumlocution nicely calculated to annoy Cope, who doubtless had "founded" the single vertebra species:

> [T]he occurrence of closely related species of large serpents in the same geological formations of Europe and America just after the total disappearance in each country of *Mosasaurus* and its allies, which show such marked ophidian affinities, is a fact of peculiar interest, in view of the not improbable origin of the former type; and the intermediate forms, which recent discoveries have led paleontologists, familiar with these groups, to confidently anticipate, will doubtless, at no distant day, reward explorations in the proper geological horizons.

Cope's enthusiasm for "ophidian affinities" burgeoned when he went west in 1871. Encountering many fossil mosasaurs in the Kansas Chalk bluffs, he excitedly attributed truly monstrous proportions to them. One, he avowed in a government report, "was probably the longest of known reptiles, and probably equal to the great finner whales of modern oceans."

"We found the bones of the head of a huge mosasaur projecting from the bank," he wrote his wife.

> We at once began to dig and soon came on the skull in the bank! We worked nearly the whole of two days, the men helping, and succeeded in getting out nearly the whole of the important part of . . . [a] very fine specimen and a rare animal. The tail went entirely through one spur of the bluff and we found the end on the other side. The animal must have been 75 feet long. We had to leave a part of it in the bank.

As with his head-to-tail plesiosaur, Edward's volatility carried him away. Apparently, he mistook the tail of a mosasaur skeleton on the bluff's "other side" for that of the "rare animal" he was excavating. The genus they had found, *Tylosaurus,* grew to fifty feet at most, rarely exceeding thirty-five. Although the rearing specimen in Zallinger's painting deeply impressed me as a child, *Tylosaurus* was definitely not the hundred-foot *Hydrarchus* of Edward's primal memory, and its limitations soon became evident. Marsh must have gloated, especially after his rival's bragging about his twenty-five-foot-wingspan pterosaur.

Still, sea serpent mania continued through Cope's life, well past the *New York Herald* showdown with Marsh. "If I had the money I would hire a small vessel and go on a cruise for him," he wrote about a reported sighting in 1888. "The chance of seeing this famous beast may never be so near again. I am very curious to know if it is a huge eel, or a reptile, or a zeuglodont cetacean of slender proportions." Returning west in 1892 and 1893, he encountered Indian legends that "fossil bones are those of huge serpents which burrow in the earth." He described them obsessively in letters to his wife, and although he professed not to be "troubled by such superstitions," an 1895 event revealed his susceptibility.

After reading in a Philadelphia morning paper that "the Sea serpent had been caught" near New York City, he jumped on a train north and spent the day dashing around Manhattan newspaper and customs offices in search of it. Ballou, ever ready to mount on Cope hobbyhorses, told him it had been seen at a taxidermist's, so they "visited two of these gentlemen and only one of them had seen it in a barrel and it was sent away, he did not know where!" The rumor had arisen from a boa constrictor or anaconda specimen, and Cope "had six minutes to catch the 3 p.m. train home, which [he] did by the skin of [his] teeth."

As death ended Cope's sea serpent quest, ironically, fossils of a diversity of serpentine marine or semimarine lizards came to light around the Mediterranean, which in the Mesozoic had been part of a great tropical seaway, the Tethys, dividing the supercontinents of Laurasia and Gondwana. The aigialosaurs, the possible mosasaur ancestors, comprised one group of relatively large such reptiles, and other, small-to-medium-sized ones called dolichosaurs, coniasaurs, and pachyophiids also seemed related to varanoid lizards. Some European paleontologists, notably a wealthy Hungarian, Franz Baron Nopcsa von Felso-Szilvas, whose career was even stormier than Cope's (including a bid for the Albanian throne), thought such groups might be related or ancestral to snakes as well as mosasaurs.

The herpetological majority eventually sat on marine-snake-origin enthusiasts, and Mediterranean sea lizards receded into prehistory. Uncrowned,

Nopcsa killed himself in 1933 after losing his money. In 1997, however, Michael W. Caldwell and Michael S. Y. Lee (the paleontologists who described aigialosaur embryos from the Balkans) revived the controversy by publishing an article about a fossil found twelve years earlier in a mid-Cretaceous marine deposit in Israel. Named *Pachyrhachis*, the three-foot creature had a serpentine shape, but small hind limbs with well-developed bones. It had been classified as a lizard, but Caldwell and Lee found "compelling evidence" in its skull and other bones that *Pachyrhachis* had been more like a snake. This again raised the possibility that snakes, like mosasaurs, had evolved from the complex of marine lizards that thronged the Tethys Sea in the early Cretaceous.

It raised the further possibility that snakes are originally marine creatures that have returned to land, a phenomenon that would be even more unusual than snakes evolving in water. In another article, Lee and Caldwell acknowledged that "considering the repeated invasion of aquatic habitats by tetrapods, and the infrequent transitions in the other direction," a land origin for snakes seemed "more plausible." After analyzing relationships among the various Cretaceous marine lizards, however, they concluded that "marine habits are primitive for pythonomorphs" and that "[the] corollary is that snakes evolved in a marine environment and are secondarily terrestrial."

Lee and Caldwell were less conclusive as to how this might have occurred, "given that fully terrestrial vertebrates have presumably evolved only once (origin of tetrapods) suggesting that an aquatic-to-terrestrial transition is unlikely and very difficult to achieve." They pointed out, however, that "while most secondarily aquatic tetrapods have paddles for swimming, and are thus clumsy on land, dolichosaurs and similar taxa, and primitive (pachyophiid) snakes possess body forms useful not only for swimming but also for burrowing, moving through crevices, and traversing open ground." They suggested that if the first terrestrial snakes were indeed burrowers in accordance with majority opinion, then perhaps "the initial colonization of land involved an aquatic (or semi-aquatic) to fossorial [burrowing] transition, without an above ground, terrestrial phase."

In other words, snakes might have evolved from underwater to underground habitats. Marine snakes like *Pachyrhachis*, for example, might have rested between swimming forays in sea caves and found subterranean life a peaceful alternative to the predatory seas. The cave on the south side of Point Resistance suggests the possibilities, a spot protected on all sides by cliffs and reefs, with holes leading into the rocks.

There was no direct fossil evidence of such a sea-to-land transition, and the idea swiftly drew a rebuttal. In a 2000 article, a group of paleontologists led by Eitan Tchernov of Hebrew University maintained that another fossil, *Haasiophis,* from the same Israeli quarry that had produced *Pachyrhachis* demonstrated that both genera were more closely related to living terrestrial snakes such as boas and pythons than to extinct marine lizards. Thus, according to their analysis, the Israeli sea snakes and the Mediterranean sea lizards had evolved from separate, terrestrial groups of varanoid lizards. Snakes were not a virtually unique instance of a major sea-evolved reptile coming ashore, but simply one of many land groups that had branched out into mainly terrestrial, but partly marine, habitats.

In 2006, the announcement of a newly discovered Cretaceous snake from Patagonia seemed to lend more support to the land origin theory. The fossil species, named *Najash rionegrina,* not only had vestigial hind limbs; it retained a sacrum, a skeletal support for the pelvis that other living and fossil snakes lack. Judging from its anatomy and the rocks wherein it was found, it was a terrestrial burrower. Still, its discovery does not completely rule out Lee's and Caldwell's intriguing scenario of sea snakes crawling into shoreline caves and learning to burrow.

Given the complexity of the marine snake controversy and the scarcity of fossil evidence on either side, the layperson must suspend judgment. Yet there would be a mythic fitness about snakes reversing one of evolution's main patterns. They are such anomalous, ambiguous creatures. Pacific rubber boas are good swimmers and bear live young, although they spend most of their lives burrowing in the forest floor. Most snakes are good swimmers—I once encountered a large rattlesnake navigating skillfully between islets in a brackish estuary. Two large families of marine snakes inhabit tropical oceans, one that gives birth, one that lays eggs. A species I've seen in Costa Rica, the yellow-bellied sea snake, sometimes occurs in large numbers off the Mexican and Central American west coasts. Both families probably evolved from land snakes, but they show that the "sea serpent" is a viable life form.

If originally marine snakes did come ashore in the late Cretaceous, it would seem emblematic of the great "revolution," as Cuvier put it, then underway. The land fossil record of the Cretaceous to Paleocene transition, as shown at a unique formation called Hell Creek in eastern Montana, demonstrates that the terrestrial snakes then living were among the groups that passed through the mass extinction almost unscathed. The appearance of huge land snakes in the Paleocene, before mammals started growing large, suggests that snakes may actually have benefited from the mass extinction. The same does not seem true of their marine relatives, since the known fossil record shows that the seagoing varanoid reptiles failed to survive the Cretaceous. So, if snakes did evolve from sea to land in the late Cretaceous, they not only performed a unique evolutionary feat but resurrected themselves from a dying old world into a living new one.

That west coast probably was not very lively, however, for millions of years after late Cretaceous events killed 70 to 85 percent of organisms. For one thing, that mass extinction saw the end of some mammals that may have been evolving an aquatic, if not marine, way of life. Marsupials called stagodonts, which left fossils at Hell Creek, seem, judging from their skeletons, to have had bulbous teeth like sea otters and flattened tails like beavers. Stagodonts were large for Mesozoic mammals, which suggests they had escaped the competition with land-dominating dinosaurs that kept most mammals rat-sized. They may have swum in estuaries, eating mollusks. But stagodonts disappeared, along with most other North American marsupials, after the Cretaceous.

A west coast seascape in the Paleocene epoch a few million years after the mass extinction would have been similar physically to the terminal Cretaceous, with cliffs and forested ridges. It's hard to know what animals would have been present, because few Paleocene fossils are known from the west coast. Turtles and birds doubtless were there, since both groups spanned the mass extinction. Still, it must have been a strangely empty world compared to the Mesozoic.

All that was about to change, however, as Professor Harlan's serpentine whale, *Basilosaurus,* showed. It lived in the Eocene epoch, which followed the Paleocene, beginning about 55 million years ago. Again, as with Triassic ichthyosaurs and Cretaceous mosasaurs, various animals would go to sea and evolve pelagic forms before land relatives developed their own special adaptations to climbing trees or eating grass. The west coast would witness a greater diversity of marine tetrapods than ever before.

SIX

Hooves into Flippers

The Paleocene and Eocene fossils Edward Cope found in the West in the 1870s gave him many ideas about how land mammals were evolving from the Mesozoic's small species. Most of the fossil beasts were very different from living groups, with primitive sets of incisor, canine, and molar teeth unlike the specialized ones of today's carnivores or ungulates, although some were getting fairly large by the Eocene. Cope usually wasn't slow to speculate about what the primeval mammals might have evolved into. When he obtained a bulbous little skull from the Wyoming Eocene, for example, he decided it was the ancestor not only of "the Malaysian genus of lemurs, *Tarsius*," but of anthropoid primates, "the family from which the true monkeys and men were derived."

As his whale-fossil collecting showed, Cope was interested in aquatic as well as land mammal evolution. When he described another Eocene fossil he'd found in Wyoming in 1872, he noted suggestive characteristics. He named it *Mesonyx*, "half claw," because its nails had a hooflike flatness, but they reminded him of something besides running. "The flat claws are a unique peculiarity, and suggest affinity to the seals, and an aquatic habitat," he wrote. "The teeth, moreover, show a tendency in the same direction, in the simplicity of their crowns." Cope speculated that *Mesonyx*'s teeth would have fitted it for preying on turtles.

Cope was uncharacteristically cautious, however, about further guesses concerning *Mesonyx*. It baffled him because the rest of its skeleton was clearly that of a land mammal. (Zallinger's *Age of Mammals* shows it as a low-slung, doglike beast.) "The structure of the ankle forbids the supposition that these animals were exclusively aquatic," he wrote, "as it is the type of the most perfect terrestrial animals." He decided that mesonychids, as he named the group, were related to land carnivores, not sea mammals. Because of their massive jaws and teeth, he thought them ancestral to hyenas.

When Cope later held forth on whale evolution, he didn't even speculate on possible links to land mammals. "The order of Cetacea is one of those of whose origin we have no definite knowledge," he wrote in 1890, confining his discussion to cetaceans' specialized necks, teeth, and ribs. He did interpret whale evolution according to his neo-Lamarckian theories, but cautiously, just suggesting that their shortened necks, reduced or nonexistent teeth, and "loss of rib heads" were "probably due to disuse."

Cope's caution on whale evolution may have reflected some rashness on his opponents' part. In the first edition of his *Origin of Species,* Darwin had leapt headlong into the yawning gulf of cetacean origins with one of scientific history's more speculative passages:

> In North America, the black bear was seen by Hearne swimming for
> hours with widely open mouth, thus catching, like a whale, insects in
> the water . . . Even in so extreme a case as this, if the supply of insects
> were constant, and if better adapted competitors did not already exist
> in the country, I can see no difficulty in a race of bears being rendered,
> by natural selection, more and more aquatic in their structure and habits,
> with larger and larger mouths, till a creature was produced as monstrous
> as a whale.

Attempting to counter one of Owen's anti-evolution arguments, he'd also cited a supposed Mesozoic whale bone as evidence against an abrupt cetacean appearance in the Eocene.

Marsh, in the same 1877 speech wherein he'd exulted over toothed birds, had asserted that whales were "connected with the marine Carnivores [that is, to seals and sea lions] through the genus *Zeuglodon,* as Huxley has shown, and the points of resemblance are so marked that the affinity cannot be doubted." Such assertions could indeed be doubted, however, and Marsh had waffled as he continued: "That the connection was a direct one, however, is hardly probable, since the diminutive brain, large number of simple teeth, and reduced limbs in the whales, all indicate them to be an old type, which doubtless branched off from the more primitive stock leading to the carnivores."

Cope must have enjoyed these flounderings. The critics had pounced on Darwin's mayfly-eating bear with such glee that he dropped it like a hot potato from later editions, and his Mesozoic whale fossil was simply a geological howler. Huxley's and Marsh's "marine carnivore" whale ancestors were less embarrassing, but still little more than a guess. Another influential evolutionist whom Cope liked to quote, William Flower, made a more convincing case for cetaceans evolving from piglike ungulates.

Cope's caution also had a more estimable side. He generally speculated on animals of which he had large collections, and in 1890 nobody had large collections of early whales. Cope's own *Basilosaurus* specimen was so fragmentary that he didn't try to name a new species from it, another instance of unusual restraint. Harlan's *Basilosaurus* and a few similar Eocene species, the earliest known whales, were unlike any known land mammal, and just as unlike any living whale. *Basilosaurus* probably looked more like a mosasaur or early ichthyosaur than a modern whale, except that its tail fins were on a horizontal rather than a vertical plane.

As it happened, however, Cope's uncharacteristic caution toward marine mammal evolution did not serve him particularly well in the long run. His conservative association of mesonychids with modern carnivores like hyenas was mistaken, for one thing. Because of their blunt teeth, later paleontologists associated them instead with a miscellaneous group of primitive ungulates that Cope had named "condylarths." And his refusal to speculate on cetacean origins proved ironically timid when the very mesonychids whose aquatic traits he'd noted but then passed over became prime suspects for whales' land forebears.

In the 1960s, a paleontologist, Leigh Van Valen, commented on a remarkable similarity between mesonychids' teeth and early whales', suggesting a relationship. In the 1970s, another paleontologist, Philip Gingerich, who had been studying mesonychids in Wyoming, decided to seek their origins in Asia, where many Eocene mammal groups were thought to have begun. Looking in Pakistan's early Eocene formations, Gingerich found the cranium of what appeared to be a mesonychid, except that, puzzlingly, it had ear bones like a whale's. Finding more of the creature (which Gingerich named *Pakicetus,* "Pakistani whale") took years, but when most of it had turned up, it became clear that, despite whalelike traits, it had been like *Mesonyx,* a semiaquatic land beast. The sediments it lay in suggested that

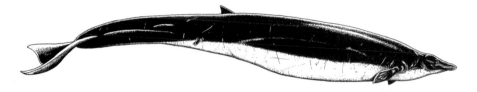

Figure 7. Despite its genus name, "ruling reptile," *Basilosaurus* was one of the earliest fully marine whales. A fossil from Vancouver Island suggests that it or a related genus inhabited the west coast 50 million years ago.

it lived by running around the margin of the dwindling Tethys Sea in search of fish and other aquatic prey.

By the 1990s, Eocene Tethys Sea sediments had yielded a diversity of whalelike mammals, rather as Cretaceous ones had yielded a diversity of snakelike lizards. There evidently had been a progression from "land whales" to sea whales, although some aquatic forms had lived earlier than some land ones. *Pakicetus* was 2 million years younger than a fossil named *Ambulocetus,* a short-legged, long-snouted "whale-crocodile" that lurked in rivers. And *Ambulocetus* was not much older than a creature named *Rhodocetus,* which, although it had four functional, if stubby, legs, had been a mainly marine creature that swam with its tail as living whales do. Gingerich discovered that even serpentine *Basilosaurus* had possessed a functioning pair of hind legs, although they were so tiny that it used them, not for locomotion, but perhaps for copulation, as boas do.

Doubt later arose as to whether whales are descended from mesonychids. Analysis of *Pakicetus*'s teeth and ankle bones, and of living whale DNA, suggested that whales are more closely related to artiodactyls, the even-toed ungulate group that includes most living hoofed animals, including William Flower's pigs. But since both artiodactyls and mesonychids originated from primitive ungulates, the distinction does not seem that important compared to the remarkable evolutionary transition from a dog-like mammal to a mosasaur-like one that *Pakicetus* and *Basilosaurus* reveal.

Parallels between whale and mosasaur evolution go beyond origins. Having evolved among a diversity of semimarine creatures in the warm, shallow Tethys Sea, both soon spread. Eocene whale fossils have turned up as far away as New Zealand and British Columbia. "The discovery on Vancouver Island of a lumbar vertebra belonging to a fairly large archaeocete furnishes conclusive proof that at least one representative of this group had arrived in the North Pacific Ocean," wrote a Smithsonian Institution biologist, Remington Kellogg, in 1936. The first pelagic whales would not have had too far to go from the Tethys to the western North American coast: the Atlantic was narrower, and a wide Central American seaway connected it to the Pacific.

The Vancouver Island archaeocete bone fits with what is known about North America's Eocene west coast. The environmental extremes of the late Cretaceous and Paleocene had moderated by 50 million years ago, and the planet was undergoing a greenhouse period. Plant fossils show that tropical forest grew as far north as Alaska along the Pacific shore, much of which was low-lying and swampy. Bird fossils from Oregon include boobies, pelican relatives common off Central America's west coast today. Early whales would not have needed later ones' blubber and other adaptations to thrive in the warm Eocene waters, feeding on abundant ray-finned fish and squid, two other groups rapidly diversifying in the vacuum left by extinction of lobe-finned fish and ammonites.

Marine tetrapod evolution in the recent, Cenozoic, era was not simply a reprise of the Mesozoic, however. Flowering plants had evolved in the Cretaceous, and mammals with teeth evolved for fine cutting and grinding quickly took advantage of the nutritious new food source. The dominant living mammal groups—marsupials and placentals—may have originated partly through feeding on angiosperms, since they appeared at about the same time. By the Eocene, angiosperms had invaded coastal waters in the form of sea grasses, which covered vast areas of shallow bay and estuary. Mammals quickly followed them, which may have been the beginning of large-scale marine herbivory by tetrapods. None of the known Mesozoic marine reptiles except possibly turtles seemed adapted to feed on marine plants.

When Richard Owen described a fossil skull and jaws from Jamaican marine sediments in 1855, he identified them rightly as belonging to a sirenian, an early relative of the herbivorous manatee and Steller's sea cow. He named it *Prorastomus,* "forward-jaw-mouth," in reference to sirenians' downward-pointing mouth. Its teeth were less specialized than recent sirenians', with canines, incisors, premolars, and molars instead of manatees' few molars or the sea cow's stumpy grinding surfaces. But it clearly was a plant eater. Its age was unclear then but eventually turned out to be early Eocene, contemporary with the first whales.

Early sirenian fossils were even scarcer than early whales, so they didn't cause as many arguments. But paleontologists at first found their origins equally puzzling. "That the Sirenians are allied to the Ungulates is now generally admitted by anatomists," said Marsh in 1877, but he failed to specify exactly which "Ungulates." Cope was even more cautious in 1890. "The derivation of the Sirenia is shrouded in mystery," he wrote. "They have evidently diverged from land mammals of primitive placental type, and have become specialized in accordance with their peculiar modes of life, and have in many respects degenerated."

The world didn't have to wait as long to find out about sirenians' origins as whales', however. When British paleontologists opened Eocene deposits in Egypt's Fayum Basin southeast of Cairo around 1900, they found yet another diversity of incipient marine animals evolving on warm Tethyan shores. In 1907, Henry Fairfield Osborn, the American Museum of Natural History's Curator of Paleontology, led an expedition to the Basin and was pleased to specify just what kind of "Ungulate" sirenians were "allied to." Knowing that sirenian fossils had turned up earlier at Eocene quarries near Cairo, Osborn located them at Fayum, compared them to other fossils there, and reported "strong evidence that the Sirenia, or sea cows, represent an aquatic offshoot from the very stock that gave rise to the elephants."

"The possible affinity of the sea-cow to the elephant was shrewdly surmised by the great French anatomist De Blainville long before the days of Darwin," Osborn wrote.

> Since his time, facts favoring his bold conjecture have been slowly accumulating, but the strongest confirmation it has yet received now lies in the resemblances between the ancient sea cows (Eotherium) and the most ancient of the elephants (Moeritherium), so numerous as to be almost conclusive of very distant cousinship. When one contrasts the existing elephant with the manatee of Florida or with the dugong of Africa, and is told of this possible community of parentage between these very antipodes of structure, one is reminded of the opening sentence of a prayer by a noted professor of logic: 'Paradoxical as it may be, O Lord, it is nevertheless true.' But in evolution all things are possible, and the nascent elephant and nascent sea-cow of Eocene Libya are by no means so far apart that a faithful Darwinian cannot conceive of possible modes of derivation of both from a common ancestral stock. They represent the extremes of fitness produced by the search for food in the grasses of the river bottoms, by the sea-cows, and of the grasses and shrubs of the plains by the elephant.

Sirenians evidently had undergone an early evolution in North Africa not unlike that of cetaceans in southern Asia, but as herbivores instead of carnivores. Beginning as medium-sized land ungulates, they had adapted to bottom-feeding by evolving unusually dense, heavy bones and mouths that opened downward instead of forward. Early species like Osborn's Fayum *Eotherium* (now *Protosiren*) and Owen's *Prorastomus* had functional hind legs and probably could move on land, like *Pakicetus* and *Ambulocetus.* The hind legs became vestigial in later ones, as with *Basilosaurus,* and, like modern whales, living sirenians have no visible hind limbs, although vestigial bones persist. Like whales, they evolved front flippers and finlike tails on a horizontal plane.

As Jamaican *Prorastomus* shows, sirenians also spread during the Eocene. Early Sirenian fossils are known from southeastern North America and southern Mexico as well, so it seems likely that, along with whales, some spread west through the Central American seaway. Even northern oceans were warmer than today's, with higher sea level and fewer cold currents. Although no Eocene sirenian fossils are known from the west coast, there are fossil sea grasses. "We would expect them to have been rather widely distributed in the early Eocene," say three sirenian experts.

If early sirenians inhabited the west coast then, coastal waters might have seemed not so unlike the late Cretaceous, with the shiny head of a *Prorastomus* relative in the shallows like a plesiosaur's while basilosaurs swam offshore like mosasaurs. There would have been sea turtles, and large early seabirds related to albatrosses and pelicans. There are fewer fossils of Eocene marine life in western North America than of Cretaceous, so this is more conjectural than a Cretaceous picture. But we do have one advantage in imagining Eocene life. Its distant relatives are still alive.

There's little fossil evidence that Mesozoic marine reptiles moved into streams, for example, but many living marine mammals return to freshwater. I've seen manatees far up Florida's St. Johns River, wintering in warm limestone springs, eating water lettuce. I've seen almost as many cetacean species a thousand miles up the Amazon as at Point Resistance. The large white Amazon River dolphin, *Inia,* and the smaller pink and gray South American river dolphin, *Sotalia,* were daily sights. *Inia,* long-snouted and blind, often sounded beside the boat at night with eerie breathing sounds. On the west coast, whales and porpoises occasionally wander into San Francisco Bay and up the Sacramento River. Assuming Eocene marine mammals took after their descendants, west coast shorelines may have seemed even stranger than Cretaceous ones. *Prorastomus*-like early sirenians might have spent part of their lives trundling around swamps and marshes. Cetaceans with functional legs might have basked on beaches and offshore rocks or lurked in rivers, perhaps rushing ashore crocodile-fashion to prey on primitive ungulate relatives.

If such a strange world did exist on the west coast, its days were numbered. The mid-to-late Eocene definitely resembled the Cretaceous in verging on drastic environmental change, and although Eocene change wouldn't extinguish groups as old as plesiosaurs, it would be as momentous in its way. Continental movements were poised to revolutionize global climate as ongoing breakups of the Gondwanan and Laurasian supercontinents redistributed polar landmasses. In the north, Laurasian fragments began to isolate the Arctic Ocean from warmer seas. In the south, a ring of deep, cold water formed around increasingly isolated Antarctica. Isolation increased water temperature gradients, with cold polar water sinking, forming bottom currents that flowed toward the equator.

Polar climates began to cool, and mountain glaciers formed on Antarctica by the mid-Eocene, although forests still covered most of that continent. Temperatures fluctuated through the late Eocene and subsequent early Oligocene epochs. Antarctic glaciers receded, then advanced again and formed an ice sheet. The general trend was toward increased cold, and even early cooling stages proved too much for many marine organisms. Some, such as reef corals, followed tropical forest southward. Some disappeared, among them basilosaurs and legged sirenians, both gone by the Eocene's end.

All this change might have been expected to cause a void of marine tetrapods like that following the Cretaceous on North America's west coast. And coastal life may have thinned for a while. But cooling had advantages for some marine organisms. Cold polar currents caused upwellings that brought nutrients from the sea bottom, fostering photosynthetic plankton and feeding an abundance of fish, mollusks, crustaceans, and other animals. Nutrient upwellings especially affected coasts, fertilizing shallow-water plants able to adapt to cooling. Mammals have ancient specializations for adapting to cold, and while cetaceans and sirenians had lost their oldest one, fur, while adapting to Tethyan waters, their descendants were developing others such as accumulation of subcutaneous fat and regulation of blood circulation. Both groups would have long careers on the west coast. And other things were beginning, even more mysterious than the cetacean and sirenian odysseys from Tethys to Pacific.

It would take a while to discover them. For years, knowledge of west coast marine evolution during the rest of the Age of Mammals was as sparse as it still is for the Paleocene and Eocene epochs. Before 1970, "the only authority in the United States" on the subject was Remington Kellogg, the

Smithsonian biologist who described the Vancouver archaeocete in 1936. A Missourian who'd gotten interested in marine mammal fossils at the University of Kansas, he made them a lifetime vocation after John C. Merriam, the Klamath ichthyosaur discoverer, invited him to prepare a report on them in 1918. His job often diverted him to more practical subjects such as World War I trench rats, however, and there just weren't many Cenozoic marine fossils to study during most of his career.

The World We Live In manifests this hiatus in marine mammal paleontology. A picture of Cenozoic marine fauna might logically have followed Zallinger's *Age of Mammals* as his Mesozoic marine painting does his *Age of Reptiles*. But scientists knew so little about west coast marine tetrapods during the past 50 million years that this was impossible. So a painting of modern sea life by another artist, Rudolf Freund, follows the *Age of Mammals* and shows only one tetrapod, a sperm whale, giving no hint as to its evolution.

On the whole, Cope and Marsh had been wise to tread cautiously on Cenozoic western shores. Indeed, when Marsh briefly let down his guard upon acquiring a particularly intriguing California fossil, it turned out to be not only different from what he thought it was, but something so utterly unexpected that Cope never got to enjoy the Yale professor's mistake. Paleontologists took over half a century after the rivals' deaths to figure out what it was, and are still not entirely sure.

SEVEN

Marsh's Deceptive Desmostylians

Elk congregate for fall rut on some Pacific beaches, and I recall in 1970 walking between a herd and surf wherein seals and harbor porpoises were feeding on a fish run. A setting sun backlit the sea mammals, emphasizing their differences from the land ones, and, years before paleontologists found walking whales, the idea that porpoises are more closely related to elk than to seals would have startled me. I still find it curious, as when I recently came upon a harbor porpoise calf beached near Point Resistance. Ravens and vultures had pecked out the eyes, but it was intact otherwise. Its little flukes, flippers, and dorsal fin seemed as uniquely shaped for the sea as the legs and hoofs of two blacktail fawns on a nearby dune did for the land.

The sheer adaptedness of extant organisms has always been a stumbling block to Darwin's idea that life has changed in an undirected way, not to fulfill a cosmic plan, but simply to accord with changing conditions. Anti-Darwinians like Owen were understandably obdurate in their belief that natural selection, what another skeptic called the "law of higgledy-piggledy," would be inadequate to turn a land animal into a sea one. Even evolutionists, including Darwin, have had a sense that there must be some cosmic impulse toward improvement. Many tended to see life as moving inexorably toward perfect adaptation, and distrusted ambiguities, as Cope did his seal-clawed, hyena-like *Mesonyx*.

"We may assign the relationships of aquatic animals in proportion to the length of time that has elapsed since their separation from their terrestrial relatives," wrote Remington Kellogg, summing up evolutionist progress-mindedness. "Also, the amount of adaptation is intimately connected with the length of time that has elapsed since they were subjected to the influence of water." In other words, any beast that spent millions of years at sea could be expected to steadily evolve things like flippers and flukes.

Othniel Marsh was progress-minded in no uncertain terms. Although a strong advocate of natural selection, he was too much the Yankee pragmatist to pay attention to its untidier implications. When he sought links between groups of organisms, he wanted straightforward direction—from toothed birds to beaked ones, from three-toed horses to one-hoofed ones. Faced with a new west coast marine fossil, he accordingly assumed that he had found a two-flippered sea ungulate evolved from a four-legged land one.

California had attracted the acquisitive Othniel since his youth, when he'd almost joined his Uncle John in the gold rush. (The uncle got rich, but employees later murdered him for his stinginess.) "I wish I could describe the coast there," a Yale colleague had written in the 1860s, "the rocks jutting into the sea, teeming with life to an extent you, who have only seen other coasts, cannot appreciate. . . . More species could be collected in one mile of that coast than in a hundred miles of the Atlantic coast." Marsh had reached the Pacific during his 1870s expeditions to the western fossil beds and doubtless had hoped something like his toothed birds might turn up there.

It took a while, but Marsh was patient. In 1888, he published a paper proclaiming a new marine mammal genus based on some teeth and vertebrae from Alameda County sandstone. The rest of the skeleton was missing, and Marsh noted that some of the teeth were odd, molars "composed of a number of vertical columns, closely pressed together . . . very distinct from anything hitherto discovered in this country." Tusklike incisors resembled those of living dugongs, however, and although no other sirenian fossils were then known from the west coast, Marsh concluded that the teeth and vertebrae had belonged to a dugonglike animal, which he named *Desmostylus hesperus*, "bound pillar of the west."

Marsh did not elaborate on exactly how or when he'd "obtained" the fossils, except that it had been "in exploring a Tertiary deposit in California a few years since," and that a Dr. L. G. Yates had found the "type specimen." But his conclusion that the animal was a two-flippered, dugonglike sirenian was reasonable, since the fossils came from the Miocene epoch, millions of years after the first sirenians appeared. Dugonglike Miocene fossils such as *Metaxytherium,* to which Marsh likened his *Desmostylus,* already were well known from Europe and the eastern United States.

Marsh had reason for reticence. Edward Cope had acquired some of those sirenian fossils, which meant he craved more, and as their enmity grew, his activities increasingly unnerved the Yale professor. Legend has it that Cope once stole a whale skeleton from Harvard's Museum of Comparative Zoology, quite an accomplishment if true. Whatever their veracity, such rumors made Marsh so secretive that his travels after the 1870s are obscure. We know how he acquired *Desmostylus,* however, because L. G. Yates later said that he had sold him the fossils when Marsh visited California in 1880. A Bay Area dentist and amateur fossil collector, Yates also supplied specimens to Cope and Leidy.

As it happened, Cope had been in California the year before, in late 1879, also seeking marine mammal specimens. There's no evidence that he knew of Yates's odd teeth or tried to obtain them, but he may have. Yates's brother actually had found them in 1876. If so, Marsh's acquisition would have frustrated his younger rival. And Cope apparently encountered another frustration when he traveled from California to Oregon and met with Thomas Condon, a naturalist who also collected for Marsh and Leidy. His diary for October 13 says that Condon showed him "bones of some large vertebrate" from a marine formation, and he would have coveted them.

"I spent two days with Prof. Condon examining his collections, etc.," Cope wrote his wife. "I formed a friendship which I hope will last long. I am going to publish some work for him." The Oregon naturalist seems to have grown leery of acquisitive eastern paleontologists, however. "He does not think Marsh has used him quite as he should—a common complaint in the West I find," Cope wrote. Condon had loaned Marsh a box of bones, which, his daughter recalled, stayed in New Haven until after the professor's death, "so that the camels and horses and other animals waited patiently for years at Yale without learning even their own names." But Marsh's rudeness seems not to have worked to Cope's advantage in this case. He did not get to publish on Condon's marine bones.

When the methodical Marsh finally published his *Desmostylus* paper, he must have been pleased to pioneer an area where Cope's efforts had failed to produce anything remotely as novel. Cope would have envied it, and probably longed to deride it, as he often did his rival's publications. But he lacked enough information about sirenian evolution to challenge Marsh's claim, and, just before the *New York Herald* debacle, he had too many other concerns to stab at every possible crack in Marsh's armor. He ignored *Desmo-*

stylus in an 1890 *American Naturalist* review of extinct sirenians. Had he lived long enough, however, later developments would have delighted him.

William Flower, the English evolutionist who associated whales with pigs, upheld Marsh's classification of *Desmostylus* as a sirenian in 1891. But doubts began to arise in 1898, when two paleontologists, S. Yoshiwara and J. Iwasaki, found a strange skull in Japan. They sent pictures to Henry Fairfield Osborn, who saw similarities to early mastodons in the flattened, elongated skull, which had two large tusklike incisors in its upper jaw and four smaller tusks on its lower. On the other hand, the skull's molars were "bound pillars" like those of Marsh's *Desmostylus,* and the fact that it had been found in "apparently marine beds" suggested it was some kind of sirenian after all. Osborn decided it was "possibly Proboscidean . . . possibly Sirenian," a conclusion consonant with his theory of a related African origin for both. "Whatever its affinities," he marveled, "this new fossil mammal is certainly most remarkable."

Desmostylians, as the creatures began to be called, remained a riddle through the next half century. John C. Merriam, as part of his west coast fossil projects, assembled a collection of their teeth ranging from southern California to northern Oregon. One of these came from Thomas Condon, who, Merriam wrote in 1911, had obtained it "from the beach of Yaquina Bay in the northern half of the Oregon Coast." In 1915, a complete skull from northern Oregon, attributed to *Desmostylus,* showed that the Japanese animal was indeed like Marsh's. From the fossils' uniform occurrence in marine beds, Merriam concluded that the genus was aquatic, and thus probably a sirenian.

Agreement on this was general, but not universal. Some paleontologists continued to suspect proboscidean affinities. One prominent Austrian, Othenio Abel, thought that desmostylians were giant, seagoing relatives of the duck-billed platypus. Their molars seemed to him like those of multituberculates, a strange early mammal group once thought ancestral to monotremes like the platypus. That implied that desmostylians laid eggs, perhaps nesting on beaches like sea turtles.

Even the majority who saw desmostylians as placental sirenians were unsure what kind they were. "The characters of the skull and dentition," Merriam wrote, "suggest that when the whole skeleton is seen this genus may be found to differ from the known groups sufficiently to make necessary its reference to a family distinct from those thus far described." Most paleontologists thought that, like living sirenians, desmostylians ate plants, but some thought their tusks and massive molars indicated a shellfish diet. "No other mammal, relative to its size," wrote one researcher, V. L. Vanderhoof, "presents to its food so much of the hardest body tissue. From these facts, is it not reasonable to suppose that the wealth of contemporary mollusks served as its staff of life?" Most thought desmostylians lived in shallow coastal water, as their fossils suggested, but some thought them more pelagic.

Whether they were mollusk or plant eaters, pelagic or coastal, most paleontologists still assumed that desmostylians had a dugonglike shape, with front flippers and a finlike tail. In 1916, a new genus found on Vancouver Island, *Cornwallius,* showed that they had not been Miocene newcomers, since it was in a deposit from the previous Oligocene epoch. Presumably, then, desmostylian evolution had paralleled that of other sirenians. "*Cornwallius* is probably structurally ancestral to *Desmostylus,*" wrote Vanderhoof in 1937, "and with the evidence at hand, we might conclude that the migration of desmostylids took place in a westerly direction. If we accept *Prorastomus* from the Jamaican Eocene as the stem lineage, then our case is further strengthened."

Teeth, skulls, and bits of other bones were still the only desmostylian fossils known in 1937, however. When an entire skeleton finally turned up on Sakhalin Island north of Japan, it was a shock. Japanese scientists announced the discovery in 1941, so war delayed its impact, but American sirenian experts reeled. If desmostylians had been dugongs, they had been very unusual ones indeed. The Sakhalin skeleton did not have flippers or a flukelike tail; it had four legs, with bones as stout as a hippopotamus's.

In his 1952 dissertation, a UC Berkeley PhD candidate, Roy Reinhart, wrote: "[C]ertain basic differences make it impossible to place the desmostylids in either the orders Sirenia or Proboscidea. The evidence presented suggests that the Desmostylids are from a paenungulate stock related to the orders Sirenia or Proboscidea but which had been divergent from them since Paleocene time." The next year, the new Doctor of Philosophy established a brand new order for the unexpected beasts. "Range from semi-amphibious forms which were capable of limited terrestrial movement to forms almost completely amphibious," he wrote. "Size and habits similar to those of hippopotamus; swampy to brackish water environment probable. . . . Evidence from stoutly built limbs suggests movement in a fore-aft direction; feet plantigrade."

From an unusual dugong, Marsh's fossil had morphed into something closer to the Eocene African *Moeritherium* that Osborn had classed as an early elephant relative. But instead of vanishing after the Eocene like

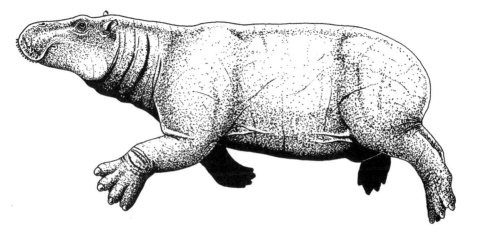

Figure 8. Desmostylians, hippo-sized ungulates, grazed sea grasses in west coast waters for millions of years, but, unlike other marine mammals, they never evolved aquatic traits such as flippers.

Moeritherium, the weird creature had continued to live around the Pacific, content to paddle along flat-footedly, grazing sea grass for millions of years after whales and dugongs had traded feet for flippers and fins.

The desmostylians' four feet rudely trampled the evolutionary web spun by decades of thinking of them as sirenians. Despite their similarities to dugongs and mastodons, they appeared unconnected either to the past or to the present, since their origins were unknown, and nothing like them lives today. Thriving in the North Pacific for some 20 million years, much longer than it took whales and sirenians to evolve flippers and flukes, they simply kept their four legs. The last desmostylians probably looked much like the first ones, and although they may have died out because they failed to adapt to marine life as fully as whales and dugongs, there was no evidence of this except the circumstance of their disappearance.

Paleontologists began trying to spin this apparent anomaly back into some kind of evolutionary web as soon as they discovered it. Some suggested that desmostylians might have been less marine than they first seemed, perhaps, like hippopotamuses, feeding mainly on land and entering water to rest. Their fossils are known only from marine formations, however, which suggests they lived mainly in water, perhaps coming ashore to rest and breed.

Some paleontologists suggested that their four legs, albeit not flippers, were nonetheless specialized for aquatic life. Two Americans, Charles Repenning and E. L. Packard, thought the legs might have been bent into flipperlike structures, as with sea lions, for swimming. A Japanese, N. Inuzaka, thought they might have been held in a sprawling posture, as with crocodiles, to provide stability in surf. Technical arguments accompany both theories, so it is hard for the layperson to evaluate them, but I find reconstructions based on them unconvincing. In pictures, at least, desmostylians just don't look functional with sea lion-like or crocodile-like limbs.

On the other hand, Daryl Domning, a sirenian expert at Howard University, thought desmostylian legs were simply like those of "certain large, slow moving terrestrial mammals, such as ground sloths," allowing them to clamber around tidal rocks, perhaps sitting up on their hind legs to forage for algae, and to swim in paddling fashion. And a *Desmostylus* skeleton photographed standing at Japan's Hokkaido University with foursquare, if slightly bent, legs looks right to me, although I've never seen a ground sloth clambering around, much less a desmostylian.

Domning's statement reflected the discovery in the early 1990s of marine ground sloths that had inhabited the late Cenozoic Peruvian coast. Like desmostylians, the sloths grazed on sea grasses and kelps in coastal water, although they were less heavy and probably used their claws to anchor them to the bottom when feeding. Like desmostylians, the sloths remained four-legged through millions of seagoing years, and died out for unknown reasons.

One Miocene desmostylian is named *Paleoparadoxia,* "old paradox," as though in honor of the group's hoary inscrutability. Desmostylians also seem paradoxical in that their apparent failure to evolve flippers and flukes may be better evidence for Darwinian natural selection than the great changes undergone by whales and sirenians. The failure implies that there is no cosmic blueprint driving change such as anti-Darwinians like Owen imagined; that if four-legged animals keep reproducing successfully in coastal waters, they will continue to have four legs, even though other four-legged marine animals such as the earliest whales and sirenians die out. If Darwinian natural selection merely requires successful reproduction from generation to generation, flippers and flukes are accidental options for marine animals, not cosmic imperatives.

Of course, desmostylians had to evolve from something, presumably a four-legged land ancestor. Discovery of another desmostylian genus in the 1970s gave a slightly better idea of that. Named *Behemotops,* "hippopotamus-looking," the Oligocene genus from the northern Oregon coast had more proboscidean-like tooth and jaw features than later genera. As Daryl Domning showed me at the Smithsonian, *Behemotops*'s cusped molars resembled those of mastodons or other land ungulates more than the odd "bound pillars" of later *Desmostylus*. (Domning thought the "bound pillars" evolved in response to the grit in a sea-grass diet.) This suggests that Desmostylians began as land ungulates that spread across Asia from Africa like early proboscideans, moving eastward into the north Pacific instead of westward through the Caribbean as V. L. Vanderhoof thought in 1937.

"The tethytherian order Desmostylia, although all its known representatives are marine mammals, is the sister group of the order Proboscidea rather than the order Sirenia," wrote Domning and two other paleontologists when they described *Behemotops* in 1986. But desmostylians may still have surprises in store: "The relationships of this clade to the Sirenia and to other paenungulates are still not known in detail from the paleontological evidence, nor is it known when the desmostylians first entered the sea and spread along the shores of the North Pacific as far as Mexico."

EIGHT

Emlong's Whale

The discovery of *Behemotops* was part of a scientific development as surprising, in its way, as the desmostylians' 20 million years of four-legged marine life. Although amateurs like the plesiosaur discoverer, Mary Anning, helped start it, paleontology's subsequent history involved increasing professionalization. Even in the laissez-faire Gilded Age, an independent scientist like Cope was at a disadvantage against someone like Marsh. Today, thickets of restrictions cover fossil formations Cope once roamed freely, and justifiably so, since an ignorant bone digger is as likely to destroy a specimen as recover it. Yet *Behemotops* showed that significant discovery can still arise from inspiration as well as academic training.

"Knowledge accumulates no faster than critical specimens of good quality come to light," wrote one of *Behemotops*'s three describers, Clayton E. Ray, a Smithsonian paleontologist. "A corollary of the principle that collecting is the mother of progress is that discovery of spectacular new materials and/or intense dedication and great productivity in collecting seem more often than not to come from the ranks of unusually gifted and motivated 'amateurs' (amateurs in the sense that they collect for love of the doing, not by training or vocation)."

Ray was referring to Douglas Emlong, the amateur who had carved *Behemotops*'s bones out of Oregon and Washington rocks. Emlong, he noted, had seen *Behemotops*'s proboscidean affinities even as he uncovered

68

its tusks and jaws. "I am sure it is a great find," the collector had written, describing it as elephant-like, and either a land mammal or a "new and very aberrant desmostylian." Finding *Behemotops* fossils at three locations over a nine-year period, Emlong rightly had suspected they were from the same new genus. By the time he collected them, he'd been finding desmostylian fossils for years, although, like Marsh, he'd originally thought they were sirenians. Ray's estimation of Emlong's abilities was striking:

> Douglas Emlong's Promethean prowess in discovery of unprecedented vertebrate fossils, alike in beds where many, few, or no collectors preceded him, is well known to specialists having personal knowledge of his activities. . . . Emlong's instant intuition of affinities, although based on unprepared specimens, virtually no literature or comparative material, and almost no formal training, proved in this case as in many others to be uncannily perceptive, and to overshadow our own more belabored conclusions.

Emlong had started collecting marine fossils right after arriving on the Oregon coast in 1956 at age fourteen, carving a circular, coiled object "filled with brilliantly shining yellow-white agate" from creek bedrock, then running home to read in the encyclopedia that it was a Cretaceous ammonite. He'd begun searching beaches and had found his first fossil bones in a place where he sometimes had to drop his tools and flee as eight-foot waves thundered in. "It would probably seem to most people that I go to a lot of useless risk and work hunting fossils," he wrote. "But there are moments of tremendous inspiration and thrill that make up for the fatigue."

Within a few years, the teenager's collection attracted local paleontologists, including Parke Snavely of the USGS and J. A. Shotwell of the University of Oregon's Museum of Natural History. Snavely and Charles Repenning, another USGS scientist, often rented cabins at Otter Rock near the Emlong home. "I think they invited him to dinner with them there at the cottage, sort of mutual admiration, I guess . . . ," his mother, Jennie, recalled. "He would talk science with them." Emlong not only talked science with the much older men; he argued it, sometimes challenging them with his observations on coastal geology. His reputation soon became national. "[H]e wrote to Remington Kellogg many times and described specimens that he had found," Jennie said. Kellogg was impressed, according to Jennie: "He would venture a guess at what the animal was and encouraged Doug a great deal."

"He worked by inspiration and succeeded spectacularly," wrote Ray, marveling that Emlong had amassed marine fossils "unprecedented in numbers, quality, variety, and novelty . . . what well-trained, methodical professional geologists and paleontologists had not been able to do in the 140 years since the discovery by the Wilkes expedition of fish and whale remains at Astoria." Emlong "never forgot a specimen or a seemingly inconsequential fragment if the vagaries of storms, tides, and shifting sand exposed an additional piece years later," and he "comprehended complex geology, obscure morphology, and abstruse relationships apparently by divination."

In 1961, Emlong's precocity got him into a *National Geographic* article on coastal Oregon. "Thunderous waves lash the coast during the winter storms, eating away at the cliffs, exposing all sorts of long buried fossils," wrote Paul Zahl, a pipe-smoking senior editor who brought his vacationing family on the assignment. "Knowing where and when to look for these remains of ancient life takes a special kind of beachcomber. Such a one was 18-year-old Douglas Emlong, whom we met at Gleneden Beach, a village some 50 miles southeast of Tillamook. Douglas wants to be a professional paleontologist, and I have no doubt that if his present enthusiasm continues, he'll make the grade."

Emlong took the Zahls home, where "sprawled in his front yard, together with scores of other marine fossils, were the remains of an ancient sea mammal. It was embedded in half a ton of sandstone when he found it, and Douglas had dug it out after many hours of painstaking work, then triumphantly carried it home in a truck." The article had photos of the fossil pinniped's "fangs and front teeth," of a desmostylian's gemlike "twin molars," and of Emlong showing nine-year-old Paul Zahl Jr. a giant fossil scallop. The latter photo struck a slightly incongruous note. In a slicked-back pompadour and a bright red shirt, Emlong looked as much like a rock 'n' roll singer as an aspiring scientist.

And when Zahl asked him "to take us beachcombing for fossils," the editor perhaps got more than he'd bargained for. Leading them to the foot of "some high cliffs" on the beach, "Douglas pointed to a spot about 15 feet above our heads."

"See that line?" he asked. I nodded, recognizing an eight-inch stratum ribboning across the massively eroded cliff face.
"Now run your eye along it," he directed.
I did, and suddenly discerned what the casual beach stroller perhaps would not have seen. Chalk-white shells by the hundreds protruded from this ancient layer of sedimentary rock. . . . Shrieks of delight came from

my daughter Eda. She had discovered a heap of loose rock teeming with bivalve, snail, and tube-worm fossils. They were so abundant that we could have filled a bushel basket.

At this point, Zahl may have envisioned his winsome preteen larking away up the "high cliffs" with the dreamy-eyed Fabian look-alike. In fact, the talented Emlong was a musician; he wrote songs and performed at local events. And he had an eye for more than geological formations. The article turned abruptly from fossils to the local agate industry, "a far more popular form of beachcombing."

If Zahl was uneasy, he was prescient. Emlong's personality came to unnerve scientists as much as his intelligence amazed them. He was given to romantic daydreams that eventually verged on hallucination. "I was once in love with a haunted woman," he later wrote,

and when I was miles away from her, after seeing her, I was suddenly struck by a strange and beautiful phenomenon, while on a beach engrossed in the beauty of nature. I felt a pang of intense emotion for the girl, and turned toward the direction in which she lived. I saw a wave of bead-like particles envelope me, like crystal stardust—white and shimmering, as if borne by a gust of psychic emotion from the girl, and carrying a feeling of age old rapture with beauty beyond description.

Progressively absorbed in such experiences, Emlong never did "make the grade" of professional science. Despite an IQ of 138 and top high-school marks, he never graduated from college or held a full-time job. Even his amateur collecting was technically deficient. Clayton Ray lamented that his brilliance was "matched by an impatience with technique and a refusal (or inability) to conform to society's standards for discipline. . . . He never learned to make a satisfactory plaster jacket, keep a proper field notebook, or follow instructions. It wasn't that he would not, rather that he could not."

Emlong's childhood had been hard. His mother, whose father had put her to work in a hotel at age ten, was a tough and resourceful but restless woman. While clerking at a Michigan nursery, she'd married her boss, who had four children from a first marriage. Douglas, her only child, had undergone a difficult breech delivery, respiratory failure, and milk allergy. "[I]t didn't seem to affect his mind any," she said. "It just affected his nerves." After his father died, she took Douglas west, where she moved from town to town in southern California, working in small businesses.

Emlong inherited her resourcefulness and restlessness. He started systematically collecting things in California: "Sometimes he'd be gone all day exploring in the desert wearing nothing but a pair of red swim trunks and a little red cap. He collected pieces of Indian pottery and shards from caves. . . . Every town that we lived in, he would go to the library and read everything he could find. He'd bring home four books at a time."

When they moved to Oregon, Jennie got a job managing a tourist gift shop that produced enough income for them to set up a more permanent home in a trailer. But they never had much financial security. After graduating from high school, Emlong went to Lewis and Clark for a few weeks, then came home and used a small legacy to build a museum for his collection on the Oregon coast. It charged fifty cents admission and also sold secondhand tools, fishing tackle, household goods, and jewelry. A sign proclaimed: "Need cash? We buy almost anything." But he preferred hunting fossils to managing them. A visitor recalled the museum as "pretty minimal," a room of fossil slabs lying on boxes, its proprietor absent. Money ran quickly through his hands. Clayton Ray described how, "with no financial support and with no prospect of it, he hired a logging salvage truck and crew to salvage a whale skull from the beach, and purchased fine specimens at exorbitant prices from tourist shops and rockhounds in order to save them for science."

The museum failed in 1966, and Emlong decided to sell his fossils. "It is my desire to clear one hundred thousand dollars after taxes in the sale of my collection," he naively wrote to Malcolm McKenna, a paleontologist at the American Museum of Natural History. "[T]his collection is a freak bonus to science which could never have been acquired except through the unusual circumstances surrounding my efforts." Another sales pitch to a millionaire private collector named Childs Frick boasted that the fossils represented "the sum of more than twenty thousand hours of hard work." It was even more naive, since Frick was already dead.

Still, museums coveted the collection. In fact, the University of Oregon's Museum of Natural History tried to confiscate it on the grounds that Emlong had flouted a law that required archaeological artifact hunters to get permission from the State Land Board. The Land Board, led by then governor Tom McCall, put a stop to this, however. "It's a matter of simple justice," McCall said. "This man should be allowed to reap the fruit of his work."

Ray, Repenning, and others had been working to acquire Emlong's fossils for the Smithsonian's National Museum of Natural History, which enjoyed relatively liberal funding from the Sputnik-generated science boom. In 1967, the Smithsonian bought the collection for $30,000, a sum that Ray, knowing Emlong, arranged to have paid in $6,000 yearly installments. It also hired him to continue collecting as an associate for its research foundation.

The Smithsonian got its money's worth. By 1967, the collection contained 590 specimens and weighed 40,000 pounds. It eventually grew to over 1,000 specimens, encompassing most major west coast organisms of the past 50 million years, with at least 15 new species and two new families of marine mammals. But figures don't express its magnitude. I spent hours leafing through Emlong's field notes at the National Museum—page after page on mammal, bird, fish, and turtle bones as well as invertebrate and plant remains from all over the west coast, most of which remain unstudied to this day. The fossils take up much of the museum's Paleobiology Department and outlying warehouses. It is uncanny, as though Emlong applied a visionary power to finding bones in the rock. Paleontologists not mystically inclined themselves feel this. "I cannot explain in a rational way how Emlong did it," Clayton Ray told me.

Mysterious as Emlong's gift was, he surpassed himself in 1964, when he retrieved the skeleton of a whale that had lived at about the same time as elephantine *Behemotops* from sandstone near Seal Rock State Park in northern Oregon. It was a find of huge historic importance—the earliest whale skeleton found on the west coast, and one of relatively few complete Oligocene cetacean skeletons in the world. It was a "missing link" as significant in its way as Marsh's toothed birds.

Whales pose somewhat the reverse evolutionary problem of desmostylians'. While the four-legged sea mammals baffled paleontologists by changing so little during 20 million years, whales bewildered them by changing so much. Indeed, the two living whale orders, the Mysticeti, or baleen whales, and Odontoceti, or toothed whales, differ so much from each other and from the Archaeoceti, extinct whales like *Basilosaurus,* that some paleontologists suspected that whales are polyphyletic—descended from more than one ancestor. They thought archaic whales might have disappeared without descendants at the Eocene's end, and that baleen and toothed whales then might have evolved separately from different kinds of land animals.

Seeing a thirty-foot gray whale near a five-foot harbor porpoise at Point Resistance does suggest separate origins. Comparing the two species' skulls and an archaic whale's does so even more. As paleontologists from Richard Owen to Philip Gingerich have observed, archaic whale skulls resemble those of the Eocene's primitive land beasts, flat-craniumed and long-snouted, with the basic mammalian dentition of molars, premolars, canines, and incisors. By comparison, baleen whale skulls look extraterrestrial—huge, toothless, triangular shapes in which eye sockets, nasal apertures, and other land mammal attributes are far from obvious. Toothed whale skulls look more earthly, but in a very strange way, with bulbous, foreshortened craniums and beaklike snouts lined with uniform teeth. It is indeed very hard to see evolutionary links among them, as Darwin demonstrated in the sixth edition of his *Origin of Species,* long after he'd lived down his mayfly-eating bears, when he tried to imagine how mysticetes had exchanged their teeth for baleen, the structures they use to strain food from the water.

Darwin described bowhead whales' oral equipment:

> The baleen consists of a row, on each side of the upper jaw, of about 300 plates or laminae, which stand close together transversely to the longer axis of the mouth. Within the main rows there are some subsidiary rows. The extremities and inner margins of all the plates are frayed into stiff bristles, which clothe the whole gigantic palate, and serve to strain or sift the water, and thus to secure the minute prey on which these great animals subsist.

He then noted an observation from St. George Mivart, one of his critics, who maintained that while natural selection would "promote" such an array of bristles once it "had attained such a size and development as to be at all useful," it was hard to see "how to obtain the beginning of such a useful development."

"In answer, it may be asked," Darwin wrote,

> why should not the early progenitors of the whales with baleen have possessed a mouth constructed something like the lamellated beak of a duck? Ducks, like whales, subsist by sifting the mud and water; and the family has sometimes been called *Criblatores,* or sifters. I hope that I may not be misconstrued into saying that the progenitors of whales did actually possess mouths lamellated like the beak of a duck. I wish only to show that it is not incredible, and that the immense plates of baleen in the Greenland whale might have been developed from such lamellae by finely graduated steps, each of service to its possessor.

Baleen whale evolution from a duck-billed ancestor, perhaps a platypus relative like Othenio Abel's monotreme desmostylian, was not an impossibility, given the sparse fossil record. Most nineteenth-century paleontologists were more cautious than Darwin, however, and tried to derive the Mysticeti from other whale groups. Cope made the reasonable suggestion that the unearthly-looking giants might have evolved from the merely strange-looking Odontoceti, which in turn might have evolved from the Archaeoceti: "The structures of the mandibular rami show the transition from such a form [an early toothed whale] to those of the right [baleen] whales. Deriving the Balaenidae [baleen whales] then from a form like that of the genus *Agorophius* [an early toothed whale] . . . we have a succession of genera in which the gingival groove and dental canal show various stages of roofing, fusion, and obliteration."

Even Cope's reasonable ideas had little significance in the vacuum of transitional whale fossils, however. Marsh showed his usual shrewdness by ignoring the subject after the sweeping generalizations of his 1877 speech, and knowledge of baleen and toothed whale origins did not increase much for the next century. Fossils from all over the world suggested that Archaeoceti had dwindled and disappeared after the Eocene, and that early versions of Mysticeti and Odontoceti had appeared in the Oligocene. But there were no obvious links between them, and despite skeletal similarities as well as differences, the possibility remained that either or both baleen and toothed whales had evolved from something other than archaic whales—that their similarities were convergent instead of phyletic.

But Douglas Emlong's Oligocene skeleton was different from all other whale fossils known in 1964. The find was so significant that Kellogg, Snavely, Shotwell, and another USGS scientist, Edward Mitchell, helped him write a highly influential paper describing and classifying it. "They complimented him on it," recalled Jennie Emlong, "and it was published by the University of Oregon, but only on a limited scale. They gave Doug a few copies of it, then they sold the rest to scientists throughout the world. He received cards from people as far away as Red China, from the head of the paleontology department of the university."

As his mother hinted, the paper's publication must have seemed a mixed blessing, since it occurred the same year, 1966, that the University of Oregon's museum tried to confiscate Emlong's collection. Still, despite some misspellings and a failure to explain the etymology of the new whale species' name, *Aetiocetus cotylalveus,* it is an impressive achievement. Emlong saw that the species was much more like a mysticete whale than an archaeocete or an odontocete. Although *Aetiocetus*'s skull superficially looked more like an archaic whale's than a modern baleen whale's, snouty and flat-craniumed, it also had significant similarities to the latter. Its lower jaw bones were unfused at the chin, a feature that helps baleen whales open their mouths wide to engulf plankton, and its upper jaws extended backward to form ridges under its eye sockets, also facilitating a wide gape by strengthening the jaw joint. Neither archaic nor toothed whale skulls have such features.

Emlong speculated that *Aetiocetus* "fed upon small fish, or other easily caught prey," and that the coincidence of "considerable abrasion" on some of its tooth crowns, and "numerous crab carapaces" in the type locality's sandstone, might have indicated a crustacean diet. "The food habits of this early cetacean could have been important to the direction of its evolutionary tendencies. For example, the consumption of easily caught prey, such as crustaceans, etc., rather than pursuit of large, fast moving prey could help to establish tendencies toward adaptation of mysticete characteristics."

"No morphological obstacles exclude this cetacean from the mysticete lineage," Emlong concluded. But there was another obstacle. As he noted in speculating on its diet, *Aetiocetus* had a full, if small, set of teeth, including cusped molars and curved canines and incisors. The teeth were unlike the more uniform ones of Odontoceti, and Emlong was sure from this and other evidence that *Aetiocetus* was not part of the toothed whale lineage. But, at that time, no baleen whales with teeth were known, so he felt bound to classify *Aetiocetus* as a relict archaeocete that had survived the demise of other archaic whales for millions of years. Most scientists tended to agree. "The presence in late Oligocene deposits of toothed whales with broad rostra suggests that such a form, in late Eocene or very early Oligocene, may have been ancestral to the Mysticeti," wrote two paleontologists in 1976.

Leigh Van Valen, who proposed whale descent from Cope's mesonychids, came to a different conclusion, however. "Emlong excludes *Aetiocetus* from the Mysticeti apparently only because of the presence of teeth," he wrote. "The ancestors of mysticetes must have had teeth, and *Aetiocetus* is in other respects similar to Mysticetes." Van Valen concluded that *Aetiocetus* was simply a baleen whale with teeth. Indeed, *Aetiocetus* may have had incipient baleen, since the presence of nutrient foramina, openings for blood vessels, in its jaws suggests that the gums extended to form plates that could have helped it catch small animals. The shrinkage of *Aetiocetus*'s teeth relative to archaic whales' also implies a small animal diet. In effect, *Aetiocetus* solved Darwin's problem of "how to obtain the beginning of" baleen without recourse to a hypothetical duck-billed ancestor.

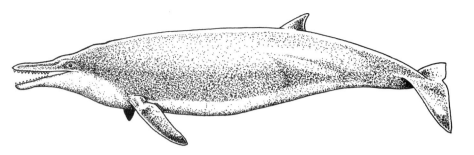

Figure 9. *Aetiocetus cotylalveus* was an early relative of living baleen whales. Unlike them, it had teeth, making it an evolutionary link between archaic and modern cetaceans.

Many other toothed mysticete fossils have been found since Emlong published his paper, upholding Van Valen's conclusion. They include an entire family of *Aetiocetus* relatives that inhabited the Pacific, and perhaps originated there. So it seems evident that baleen whales evolved from archaic ones of some kind, although the exact archaeocete group remains unknown. It wasn't *Aetiocetus* and its relatives, because even older, toothless whales with real baleen have since been found. Indeed, James Goedert and Gail Goedert, collectors who, starting in the 1970s, amassed west coast marine fossils rivaling Emlong's, found a toothless, baleen-bearing whale in late Eocene sediments on Washington's Olympic Peninsula in the 1990s. It showed that the transition must have occurred much earlier than *Aetiocetus* and, again, hinted at what a strange place the early Cenozoic west coast must have been.

Although Emlong's paper didn't perceive its full implications, his *Aetiocetus* was, like *Hesperornis* and *Pakicetus,* another of the discoveries that, as Marsh crowed in 1877, form "the stepping stones by which the evolutionist of today leads the doubting brother across the shallow remnant of the gulf, once thought impassable," between apparently unconnected groups of organisms. Emlong did what Marsh and Cope would have liked to do—he revolutionized cetacean paleontology by finding a better fossil than anyone else.

It has taken longer for any kind of a stepping-stone across the gulf from archaic to toothed whales to surface on the west coast, perhaps because the toothed whales have a distinguishing feature even more complex than baleen. All whales differ from land mammals in having ear bones encased in structures attached to their jaws only by a flange of bone. This lets them perceive the sources of underwater sounds, which pass through the jaw-contained ear bones of land mammals and thus seem to come from all directions. But odontocetes differ from other whales in that they can also echolocate—perceive prey or other objects by emitting high-pitched sounds and listening to the echoes that bounce off them. A complex system of nasal passages and valves in their bulbous, foreshortened skulls produces the sounds by recirculating air. A melon-shaped fat deposit in their foreheads allows them to direct and focus the sounds, and fat in the jaws picks up echoes and transfers them to the ear bones.

Emlong found a number of late Oligocene odontocetes, including weird ones like *Simocetus,* "pug-nosed whale," which he dug from the banks of the Yaquina River in 1977. Ewan Fordyce, a New Zealand paleontologist who described it, found its teeth and jaw different from any other known whale's and thought it might have been "a bottom feeder that preyed through suction feeding on soft-bodied invertebrates." But Emlong never found an early toothed whale as significantly transitional as *Aetiocetus* was a baleen one. "Features of the face and basicranium point to echolocation abilities comparable to those of extant Odontoceti," wrote Fordyce of the ancient *Simocetus.* There had to have been a long jump from Eocene *Basilosaurus* to such complicated creatures.

In the early 1990s, however, the Goederts retrieved another primeval fossil—the nearly complete skull of a late Eocene or early Oligocene odontocete from the Washington coast. Its cranium was surprisingly advanced, although its teeth were still somewhat archaic. The skull resembled that of *Agorophius pygmaeus,* the primitive toothed whale that Cope had described, but was actually more odontocete-like, wider and shorter. "Its cranial characters are more derived than those of many 'typical' late Oligocene odontocetes, including *Agorophius pygmaeus,*" wrote James Goedert and Lawrence Barnes, a paleontologist at the Los Angeles County Museum, "and indicates that the suborder Odontoceti must have originated at least in late Eocene times. This is much earlier than previously expected."

If the earliest known odontocete fossils haven't completely bridged the "shallow remnant of the gulf" between them and archaeocetes, a phyletic relationship seems increasingly likely. And earlier fossils of both mysticete and odontocete links with archaeocete whales doubtless will turn up eventually. "My opinion," Barnes told me, "is that we're on the threshold of finding the origins of both of them."

It still seems surprising that such sophisticated aquatic adaptations as baleen and echolocation had already evolved when the hippolike desmo-stylians were beginning their 20-million-year amble around the Pacific Rim. Yet there may be at least one Darwinian link between the west coast appearance of desmostylians and that of modern whales. Both may have been responses to environmental change. Douglas Emlong provided more possible evidence of this as he was digging up *Behemotops* and *Aetiocetus.*

In the late Eocene and early Oligocene, as increasing polar isolation cooled high latitude oceans, resultant stronger currents caused vast nutri-ent upwellings. Marine food resources would have increased because of this, but not in forms that most Eocene sea mammals were adapted to exploit. Cooling climate and resultant falling sea levels would have stressed sireni-ans used to warm, quiet bays full of sea grasses. More cold-adapted desmo-stylians, able to graze in shallow tidal zones or on offshore reefs, could have thrived, however. Archaic whales used to hunting large prey by sight in clear tropical waters would have had trouble as coasts grew cooler and more tur-bulent. Nutrient-rich cold water supports greater biomass than tropical, but much of it is small, and water turbidity tends to obscure it. The evolution of baleen to catch increasingly abundant small prey, and of echolocation to find decreasingly visible larger prey, could have helped the two newly evolved whale orders to survive the archaic ones.

Desmostylians, odontocetes, and mysticetes weren't the only new organ-isms to appear as climate changed. In 1969, Hildegarde Howard, an orni-thologist at the Los Angeles County Museum, made an daring classification from a piece of wing bone found in a mid-Tertiary deposit in the hills of Kern County, near Bakersfield. Howard thought the bone came from a member of the pelecaniformes, the ancient order that includes pelicans and cormorants, but that it was different enough from known forms to justify creating a new family, which she named the Plotopteridae, "the swimming winged." She guessed from the bone's shape that the cormorant-sized bird had been flightless, using its wings as flippers. "[T]he bone shows a basic relationship to the coracoid in cormorants," she wrote. "Other features of the bone, however, show modifications paralleling those found in pen-guins and auks."

Howard named the bird *Plotopterum joaquinensis,* but her fragmentary find got little attention. Storrs Olson, a Smithsonian ornithologist, was skeptical of her designation of a whole new family from a single bone. He was also skeptical of bird fossils Emlong was finding, but, again, the eccen-tric amateur's work would prove surprisingly significant. "Storrs Olson was bowled over by the way you called the shots on these bird specimens," Ray

Figure 10. Related to living cormorants, plotopterids, "swimming wings," were penguin-sized flightless birds that inhabited the west coast 25 million years ago.

wrote Emlong in 1974. "Recently he went back over the list and pulled out some of the apparently iffy or poor specimens that he had passed over the first time through (some because he thought they weren't bird)—we've got most of them prepared, and you haven't missed yet!"

In 1977, Emlong retrieved most of a bird skeleton from late Oligocene coastal rocks in Washington that, according to Olson, "exceeded in size any of the living penguins" except the largest. Its rigid, paddlelike wings were powerful enough to "fly" through the water penguin-fashion, and its sturdy leg bones suggested that it too had waddled about on land. But it wasn't a penguin. In fact, its wing bone turned out to be like *Plotopterum*'s, proving that Howard's new family had existed, and Olson accordingly named it *Tonsala hildegardae*. Other skeletal aspects upheld Howard's idea that *Tonsala* was a pelican and cormorant relative, although its affinities were more with the freshwater anhinga, which uses feet instead of wings for underwater propulsion. *Tonsala* must have been an even more efficient underwater predator than its closest living relative and, given its size, doubtless consumed vast quantities of fish, squid, and other prey.

Four more plotopterid species later turned up from Japan, and some were even larger, over six feet "from bill tip to tail tip," as Olson wrote. Birds of similar size probably inhabited the west coast too. Such penguinlike birds seem strange away from the polar ice most penguins inhabit today. But Oligocene poles were warmer than today's, and even larger penguins then inhabited a still-forested Antarctic coast.

West coast land fossils indicate that a lush forest still bordered the Pacific, but a less tropical one than the Eocene's. The Oligocene Weaverville flora in northern California's Trinity Alps (a few hours' drive from the Klamath limestone's Triassic ichthyosaurs) includes bald cypress and deciduous hardwoods, suggesting that the coast was like today's southeast, with large estuarine wetlands and long, level beaches. Another fossil assemblage that Emlong found on the Oregon coast, the Yaquina flora, includes maples, sycamores, dogwoods, smilax, and dawn redwoods. Giant penguinlike birds certainly seem odd in such a setting, but food evidently was plentiful, and, as Storrs Olson observes, there must have been offshore islands where they could nest safe from land predators.

With plotopteriformes, desmostylians, and the rapidly diversifying baleen and toothed whales, the Oligocene North Pacific must have been a very different place from previous epochs. And other new creatures swam in the surf. Fossils from that time indicate developments that would bring about an unprecedented sea mammal diversity, including not only beasts from tropical edens like the Tethys Basin, but ones perhaps indigenous to North America's west coast. Douglas Emlong would make central contributions to their discovery as well.

NINE

Paws into Flippers

When I first came to California in December 1968, I stayed in a cottage across Tomales Bay from Point Reyes. After living in Manhattan for the past year, I felt I'd entered a primeval world. Every morning a line of flying cormorants extended from one end of the bay to the other, and when I took out a rowboat one afternoon, I saw something dreamlike. Just before sunset, some California sea lions appeared nearby, evidently diving for herring. They let me row close enough to smell their fishy breath and sometimes surfaced around the boat. As the sun neared Inverness Ridge, they stopped diving and gathered, lying on their backs with their front flippers raised in an attitude that seemed prayerful. When it touched the horizon, they barked in unison until it had set, then closed their eyes and floated quietly while the sky darkened.

I've never seen wild animals do anything so like a religious ritual. It reminded me of something I'd read about the west coast of Ireland, where funeral processions were sometimes by boat. Gray seals were said to follow the coffins of people who had been kind to them, and to pull out near the burial to mourn. Seals can make humanlike sounds. It was said that they were human once, and that they sometimes took off their skins and came ashore again.

Such things have given seals and sea lions a particular hold on the imagination. They seem the most accessible of marine mammals, more like land animals than cetaceans or sirenians, which often have been regarded as fish. Their round heads, soft eyes, and playful curiosity are attractively humanoid.

Their shared qualities tend to confuse seals and sea lions in people's minds, as when we call circus sea lions "performing seals." They are not the same, however. Gray seals, *Haliochoerus gripus,* don't live in the Pacific, for one thing, and California sea lions, *Zalophus californianus,* don't live in the Atlantic. Unlike sea lions, seals can't bend their hind limbs forward to help them move on land (or perform circus antics), and lack external ears or "pinnae." Indeed, despite their similarities, seals and sea lions may be quite different kinds of creatures, although both are called pinnipeds, "fin feet." Their relationship is another of marine tetrapods' evolutionary riddles.

Despite pinnipeds' familiarity, their origins have been even more mysterious than cetacean or sirenian ones. So little was known about fossil pinnipeds in O. C. Marsh's time that he left them out of his 1877 speech except for his vague reference to "the marine carnivores" being connected to whales "through the genus *Zeuglodon,* as Huxley has shown." Marsh doubtless was referring to pinnipeds, since they resemble land carnivores, the order including dogs, cats, bears, and otters. Huxley had asserted, from analysis of modern sea lion and seal skeletons, that both were descended from early, bearlike carnivores. Given the scarcity of their fossils at the time, Marsh's conclusion that "the points of resemblance are so marked that the affinity can not be doubted" would have been hard to contradict.

Cope, stung by a Huxley snub at an 1878 British Association meeting, would have liked to contradict his assertion about pinniped ancestry and the conclusion Marsh drew from it. If he could have acquired enough pinniped fossils, he might have tried. But, as with the marine animal bones Condon tantalizingly withheld in 1879, they eluded him. His more than 1,400 papers contain even less on pinnipeds than on sirenians and whales.

Cope's disruptive influence had a way of outliving him, however, as with William H. Ballou's sea serpent articles. Another Cope henchman, more reputable than Ballou, was a westerner named Jacob L. Wortman, who had started collecting for him as a University of Oregon student in the 1870s. Wortman moved east in the 1880s and took a medical degree while serving as Cope's main collector, then went to work for Osborn at the American Museum in the 1890s. As brilliant and prickly as Cope and Marsh, but lacking their wealth, Wortman sooner or later quarreled with every employer. After a 1903 "retirement" from cataloguing the recently deceased Marsh's mammal fossils at Yale, he moved back west. There he fulfilled Cope's ambition by getting his hands on a sea mammal fossil owned by Thomas

Condon, a "fairly well preserved skull . . . from the Marine Miocene of the Oregon Coast." The fossil seemed a pinniped missing link, and Wortman thought it contradicted the paleontological bigwigs.

"Professor Condon has kindly permitted me to make a careful study of this unique specimen, and I do not hesitate to pronounce it easily the most important find that has yet been made in this group," Wortman wrote in 1911. "As far as I am aware the specimen represents an entirely new and hitherto unknown genus, intermediate in many respects between the sea lions and seals, with perhaps the most pronounced affinities to the latter, and at the same time exhibiting a number of primitive or ancestral characters not found in the skeleton of any known pinniped."

Wortman acknowledged that Huxley, Osborn, and almost everyone else thought pinnipeds had evolved from bearlike carnivores. He thought Condon's fossil, named *Desmatophoca,* "aquatic seal," was too early to be descended from such animals, however. "In the fossil seal before us we have a very distinct pinniped," he wrote, "exhibiting no approach whatever to the dog or bear groups." Wortman saw more similarity between *Desmatophoca* and the creodonts, the "flesh-toothed"—early land mammals that Cope had described, largely from western fossils Wortman collected.

Although they had meat-shearing cheek teeth somewhat like living carnivores', creodonts were a separate, primitive group that vanished in the Miocene epoch. In some ways, they resembled mesonychids, the early, flesh-eating ungulates and putative whale ancestors. Indeed, early paleontologists thought mesonychids *were* creodonts*,* so Cope's observations about the seal-like aspects of his 1872 Wyoming fossil, *Mesonyx,* probably influenced Wortman's thinking. He thought *Desmatophoca*'s teeth, jaws, and cranium were more like those of a creodont named *Oxyaena* than a bear's, and concluded that "creodont ancestry" was "the only possible or logical solution to the problem" of pinniped origins.

Wortman's unconventional interpretation did have logic. If whales and sirenians had evolved in the Eocene epoch from early groups, then pinnipeds, the other main living sea mammals, might be expected to have a similarly ancient origin. His creodont origin theory long remained unfalsifiable, anyway, for lack of earlier pinniped fossils.

But the west coast eventually produced a pinniped fossil older than *Desmatophoca.* In 1973, Edward Mitchell and Richard Tedford, a paleontologist at the American Museum of Natural History, identified a new genus from the late Oligocene-early Miocene Pyramid Hill area in the hills near Bakersfield, California. Naming it *Enaliarctos,* "sea bear," they described it as a "medium sized arctoid carnivore, a transitional species that departed in structure from terrestrial ursids (Hemicyoninae) and evolved in the direction of aquatic pinnipeds." The fossils consisted only of two partial skulls and other cranial material, but Mitchell and Tedford saw them as decidedly bearlike, although with less olfactory capacity and more uniform teeth than bears'. A creature chasing prey underwater doesn't need a sensitive nose to locate it or a full set of incisors, canines, and molars to seize and chew it.

"The invasion of a radically new adaptive zone by any group of organisms is a phenomenon of great interest to evolutionists," Mitchell and Tedford wrote, "yet it is rare that such major changes can be followed clearly in the fossil record." The *Enaliarctos* skulls seemed to be just such a rarity, however, and to show that a bearlike or raccoonlike mammal had gone to sea, not a creodont. The creature thus probably had done so not in the Eocene, when primitive groups prevailed, but later, when "arctoid carnivores" did. Indeed, the presence on the North American west coast of this earliest known pinniped fossil suggested that the group had invaded "a radically new adaptive zone" not in the tropical Tethys Sea's Edenic setting, but on the west coast. "The area of origin and the center of dispersal of *Enaliarctos mealsi* are not known," wrote the authors, "but we infer that the North Pacific basin is the most likely site."

The sediments entombing *Enaliarctos* had been the bottom of a shallow bay or inland sea, perhaps bordered by swamps and low, wooded hills. It is not hard to imagine an early relative of bears and raccoons taking to the water in that environment. Judging from its skull, all that was available in 1973, *Enaliarctos* was about five feet long. It had larger eyes and smaller nasal sinuses than a bear, and these specializations would have helped it in diving after fish. But its molars were still adapted for shearing meat, unlike living pinnipeds', which mainly eat animals small enough to swallow with limited chewing. *Enaliarctos*'s shearing molars suggested that it dismembered prey while eating it and thus may have brought food ashore, like a bear with a salmon.

Such features implied a coastal rather than an open ocean, or pelagic, habitat, as did an absence of sexual dimorphism in the *Enaliarctos* specimens. Male sea lions and elephant seals are larger than females because they breed in rookeries, where the biggest get the most females and pass their size on to male offspring. Rookery breeding generally is a sign of pelagic life in pinnipeds, because animals that cover large expanses of ocean in search of food tend to resort to specific locations to find the opposite sex. On the other hand, animals that spend their lives dispersed along relatively small

Figure 11. *Enaliarctos,* "sea bear," was the earliest known pinniped. It probably evolved from west coast land ancestors over 25 million years ago.

stretches of coastline undergo less competition for mates, and the sexes tend to be similar in size. So, again, *Enaliarctos* perhaps had lived more like an aquatic bear than a sea lion, coupling seasonally but otherwise being relatively solitary.

Enaliarctos more than *Desmotophoca* thus seemed another of Marsh's "stepping stones," a fossil link between living pinnipeds and Oligocene carnivores, not Eocene creodonts. But there was a catch. Its skull clearly resembled those of sea lions and fur seals, which are classed in the pinniped group called the Otariidae, and also those of walruses, the Odobenidae. But it seemed less like skulls of the other living pinniped group, the Phocidae, which includes harbor seals and elephant seals as well as Atlantic gray seals. "There are few important features of the cranium that suggest affinity with the Phocidae," Mitchell and Tedford wrote of *Enaliarctos;* "much of the evidence is in the otariid direction. . . . We conclude that we have demonstrated a sequence from hemicyonines to otariids that leaves little room for phocid origins."

But if *Enaliarctos* wasn't related to harbor seals and elephant seals, what kind of creature had phocids like them originated from? In 1911, Wortman had considered Condon's *Desmatophoca* more like phocids than otariids. But *Desmatophoca,* with nonshearing molars, was a more seagoing animal than *Enaliarctos,* less of a link between land mammals and marine ones. Anyway, other paleontologists considered *Desmatophoca* more like otariids, while Mitchell and Tedford also raised a third alternative: "We recognize the possibility that desmatophocines and their close allies might represent yet a third major pinniped group independently derived from the middle Tertiary arctoid adaptive radiation."

The idea that phocid seals originated separately from other pinnipeds was not a new one. One of the first to have it was St. George Mivart, who frustrated Darwin by doubting natural selection's ability to evolve whale baleen. Mivart thought phocids' anatomy resembled otters' more than bears', and speculated that they had originated from otterlike animals, mustelids, at about the same time as otariids and odobenids had originated from bearlike ones, ursids. A largely complete skeleton of an early Miocene creature named *Potamotherium,* "river beast," seemed to support Mivart. Common in Europe and eastern North America, *Potamotherium* was more aquatic than living otters and had a number of skeletal features similar to those of phocid seals as well as mustelids.

"It seems to me that several of the presumably derived features of the cranium and forelimb of *Potamotherium* agree specifically with the mustelids as traditionally believed," wrote Richard Tedford in 1976. "A phyletic relationship to the phocids also seems implied by these comparisons and it might be possible to regard *Potamotherium* as occupying an intermediate structural position between terrestrial mustelids and pelagic phocids in much the same way as does *Enaliarctos* with respect to the ursids and otarioids."

Other factors seemed to support a separate phocid origin. Bones of early phocids mainly came from the North Atlantic and Mediterranean. "The shores of the North Atlantic Ocean . . . have been the site of the origin, the longest part of the discovered history, and the reservoir of successive dispersal and radiations of phocids," wrote Clayton Ray in 1976. Ray thought that phocids had been confined to the Atlantic in the Miocene and had reached the Pacific later, with an early, warm water group, the monachines, entering via the Panama seaway, and a later, cold water group, the phocines, entering via the Arctic. So, presumably, phocid pinnipeds had originated at about the same time as otariid ones, and perhaps in the same way, by a land carnivore entering the water to take advantage of burgeoning food sources. But phocids had originated from otterlike animals, perhaps in east coast forests, while otarioids—sea lions, fur seals, and walruses—had originated from west coast bearlike ones.

This diphyletic interpretation of pinniped origins can seem intuitively right. When I compare sea lions and harbor seals on the rocks at Point Reyes, the sea lions do seem ursine as they slouch clownishly around, waving their necks and sparring with their teeth. The harbor seals do seem otterlike, curious and playful, but more self-contained. Still, judging from the present is deceptive, given the depth of geological time. Neither species is much like a modern bear or otter. Elephant seals, also visible at Point Reyes, are even less so.

There were stumbling blocks to the diphyletic interpretation. Despite otterlike features, *Potamotherium* wasn't an unequivocal otter relative. Some anatomical features made it seem more bearlike. And Tedford noted in his 1976 paper that the diphyletic interpretation contrasted "with that provided by the immunological comparison of serum albumins and transferrins" among living pinnipeds, which suggested "a more direct sister-group relationship of the pinniped groups." If phocids and otarioids were more similar genetically than their anatomy implied, perhaps the anatomical differences were less basic than they seemed.

Still, the published fossil record in 1976 did not support a single origin for sea lions, walruses, and seals. Known *Enaliarctos* fossils did not even support the possibility, however likely it seemed, that it had been a "fin-foot," with flipperlike limbs adapted for swimming. More complete fossils might show that it had limbs more like a bear's. *Enaliarctos* remained mysterious.

Shortly before Tedford and Ray published their 1976 papers, however, Douglas Emlong had visited Pyramid Hill. "He made many trips to California," his mother recalled, "and he collected many significant fossils from various places, but particularly the area around Bakersfield. He would obtain permission from the landowners to explore those dry canyons, and worked in exceeding heat to dig those fossils out of the area." It was an audacious move for an amateur, invading territory already worked by local professionals. "I will head [sic] your advice about the California areas, as I don't want to make any trouble for anybody," Emlong wrote Ray placatingly in 1974. Perhaps it was unwise, given his problems.

As *National Geographic* editor Paul Zahl seems to have sensed, Emlong was disaster-prone. On one youthful foray, he fell down a mountainside and broke an arm. "He banged his head on that great tree where he finally landed," Jennie said, ". . . and that stopped his fall. I've often wondered if he got some damage from that that might have affected his nerves." She thought it might have aggravated inherited epileptic tendencies. A later leg fracture had to be pinned in thirteen places. "Though he recovered fairly soon, it seemed to be bothering him more and more each year. It got so that it sort of bent—the foot and leg were bent, and he would stumble some from it."

Rugged conditions contributed to the disasters. "As it rains almost all the time during the winter fossil-hunting time, I become soaked to the skin and work that wall all day in driving rain and wind," Emlong wrote. "At times when I have carried many 50 pound pieces of material up a steep bank, I become so fatigued that I become dizzy. My arms and legs shake, the muscles jerk, and heart pounds painfully." He sometimes had to travel to fossil sites by bus, and when he had a car, often lacked money to run it. Once he had to trade the hammer he'd used to collect "the skull and partial skeleton of a giant pinniped" for enough gas to drive home. But Emlong's strange personality also played its part. Jennie said that almost all his childhood reading had been "concerned with volcanoes and earthquakes and all kinds of natural disasters."

Worse things than steep slopes and "exceeding heat" lurked in the dry canyons around Bakersfield. Emlong apparently caught valley fever, a disease caused by spores of a soil fungus, particularly common there, which can disable or kill susceptible individuals:

> At first he thought it was pneumonia and he had a bad time getting back home. The one time, anyway, he could only drive about fifty miles, and then he would have to check into a motel and rest. Then he would try a little ways the next day until he got home. He went to the doctor here, and they just gave him something for pneumonia, but they never took the proper tests to pinpoint the valley fever. . . . He was in bed a long time.

Emlong's travails didn't stop him from extracting something as extraordinary as *Aetiocetus* from the hills near Bakersfield. On April 15, 1975, he found a pinniped skeleton "in place, or nearly so, at the summit of a small hill visible from Pyramid Hill road, through a distinct gap in the hills." His gift did not always extend to instant identification, and his field notes described it as the "magnificently preserved nearly complete skeleton of a lower Miocene sea lion." It was a late Oligocene *Enaliarctos,* however, another classic link that stands today in the National Museum of Natural History's Paleobiology Department like a monument to evolutionary change, since even a layperson can recognize it as a creature like, but also strangely unlike, living pinnipeds. The public exhibits downstairs feature a replica.

Emlong had to cut the sandstone slab containing the skeleton into four pieces so he could load it into his battered Toyota, drive it back to Oregon, and ship it to Washington, D.C. But the specimen was indeed "magnificently preserved," and its scientific significance was immense. After the many years required to prepare and study it, the complete *Enaliarctos* made Clayton Ray, for one, think twice about pinniped origins.

In 1989, Ray coauthored a paper in *Science* with two other paleontologists, Annalisa Berta of San Diego State University and André Wyss, then of the American Museum of Natural History, which challenged the diphyletic interpretation of pinniped origins. The paper drew on Emlong's skeleton and a 1987 study by Wyss to propose that *Enaliarctos* was related to phocid seals as well as sea lions and walruses: "Recently, compelling osteological evidence was used to support a single, monophyletic origin of pinnipeds, a view consistent with certain cytogenetic and biomolecular evidence. *Enaliarctos,* in this latter arrangement, is hypothesized as the sister taxon of all other pinnipeds, a key position in the higher level phylogeny of the group."

Emlong's skeleton showed that *Enaliarctos's* legs were indeed flippers of a sort, making it a genuine pinniped, a "fin foot." Although more suited to land locomotion than later pinnipeds', their anatomy was enough like those of all other groups to suggest that they could have evolved into phocid limbs as well as otariid and odobenid ones. *Enaliarctos* evidently had used all four limbs as well as spinal undulations to propel it through the water, a pattern that could have evolved into both phocid (hind limbs and spinal undulations) and otariid (mainly forelimbs) swimming methods. "A large number of pinniped skeletal specializations . . . do not resemble those seen in other aquatic mammals and can only be reasonably interpreted as evidence of common ancestry. It is highly implausible to attribute these resemblances to convergence, given the differences in locomoter behavior described above."

Cope's and Marsh's contentious spirit lingered in the matter of pinniped evolution, however. Charles Repenning, one of the scientists who had entertained young Emlong at the Otter Creek cabins, promptly objected that pinniped swimming specializations *could* be convergent, since other marine mammals such as whales, sirenians, and sea otters have some similar traits. Although Repenning agreed that all pinnipeds probably had evolved from bearlike, not otterlike, animals, he maintained that this did not "automatically indicate that *Enaliarctos* and the Enaliarctidae have morphologic characters that are intermediate between terrestrial arctoids and all later pinnipeds." He insisted that "all shared derived characters of *Enaliarctos*" indicated affinity with North Pacific otariids, but very few with North Atlantic phocids. He thus referred all pinniped origins back to some earlier animal: "The common ancestor, sister taxon of all other pinnipeds, is still unrecognized, and we have no way of knowing whether it was a marine or terrestrial carnivore."

Berta and Wyss replied that pinniped swimming specializations *were* different enough from other marine tetrapods' to indicate common ancestry rather than convergence:

> While it is true that a "short, robust humerus" [upper arm bone] occurs also in cetaceans and ichthyosaurs, as Repenning points out, this characterization disregards the fact that pinniped humeri are otherwise highly distinctive and are not easily confused with those of the other two groups. Considered in concert, however, the constellation of derived features of the pinniped flipper as well as a larger number of cranial synapomorphies occur nowhere else among vertebrates and indicate common heritage.

Others beside Repenning continued to support a diphyletic hypothesis, however. In 1997, Irina Koretsky, a paleontologist then at the Smithsonian, saw distinct phocid characters in late Oligocene fragments of thigh bone found in South Carolina, fragments possibly predating the earliest *Enaliarctos* fossils. If phocids already lived in the Atlantic when *Enaliarctos* inhabited the Pacific, then they perhaps did have a different "sister taxon." Berta acknowledged Koretsky's Oligocene fossil in a 1999 textbook, calling it "the oldest known phocid and pinniped." In a 2001 publication, however, she and other authors seemed to overlook the South Carolina thigh bones, stating that "phocids, though postulated to have a North Pacific origin, are first known from the middle Miocene of the Atlantic Ocean."

In 2003, Koretsky and Lawrence Barnes reiterated the diphyletic hypothesis:

> Phocids and otarioids [sea lions and walruses] now seem to have equally early origins in the late Oligocene. But the paleontological and biogeographical evidences suggest that they have separate ancestries, with phocids arising in the North Atlantic, being most closely related to musteloids, whereas otarioids arose in the North Pacific and are most closely related to ursids. Anatomical and morphological review of characters that have been used by supporters of monophyletic origins reveals that they fail to provide convincing evidence for 'pinniped' monophyly. . . . Numerous features of the pinniped skeletal morphology strongly support a dual origin for the group.

Resolution of the pinniped origin controversy seems distant for the present. "I still instinctively find diphyly more attractive," Clayton Ray told me in 2005. "But I don't think the last word has been said. We've beaten the subject to death until we get a lot more fossils. If we had some really fine material from South Carolina it might help us."

Early North Atlantic pinniped fossils have been elusive, however. Maybe Emlong would have found significant ones if he'd collected on the east coast. The Smithsonian paid his way to do so around Washington in 1974. The Calvert Cliffs in Maryland, an hour's drive from the capital, are among the east coast's important Miocene marine deposits and must have seemed a natural subject for Emlong's powers. But he stayed east only two weeks "because he was terrified of the bustle." He also sulked because a waitress who had charmed him at a Maryland restaurant ignored him on his next visit there, although Ray guessed she'd done so simply because he sat outside her serving station. Emlong did find "important fossils . . . that other scientists had overlooked," spotting some from a Potomac River bridge, but not of seals.

Of course, there are places beside the coasts of North America to look for early pinniped fossils. If all pinnipeds originated in the North Pacific, for example, how did phocid seals get to the Atlantic in the Oligocene? The Panama seaway was open then, but Central America has produced no early pinniped fossils to indicate such a migration. The Arctic could have been another route, given most pinnipeds' preference for cold water, but a land bridge occupied the Bering Strait for much of the Oligocene, and no early pinniped fossils are known from the Arctic Basin either. Maybe Emlong would have found some if he'd collected there or in the Caribbean. But he had enough trouble getting to California.

Whatever its relationship to living seals and sea lions, *Enaliarctos* was well adapted to the late Oligocene and early Miocene epochs. It lasted from 25 million to 18 million years ago, a long time for a mammal genus, and diverged into several species that must have enlivened lonely stretches of beach and estuary. Nothing quite like it had existed before—early sirenians and whales were less pliant creatures. Its sleek body and agile limbs imply a pinniped playfulness. It perhaps swam rings around ponderous early desmostylians like *Behemotops* and teased sea turtles and penguinlike plotopterids, as sea lions tease turtles and birds.

Relations with some early Miocene creatures may have been less playful. The Pyramid Hill area that held Emlong's *Enaliarctos* skeleton produced bones of at least seven toothed whale species, a great increase over the earlier west coast fossil record. Most were relatively small, but one, according to Lawrence Barnes, was "a huge eurhinodelphid which may have attained the size of Recent adult killer whales." It had a skull over five feet long and

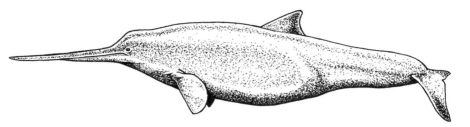

Figure 12. The genus *Eurhinodelphis,* illustrated here, was one of many long-beaked toothed whales common in the early Miocene epoch. *Macrodelphinus,* another genus in the same family, grew nearly as large as living orcas and might have been a top predator.

teeth the size of human thumbs. If this beast, *Macrodelphinus kelloggi,* had a taste for mammals, it could have played the top predator role of modern orcas. Adult desmostylians probably had little to fear from it, but their young might have been at risk, and the enaliarctines, little larger than modern sea otters, would have been bite-sized morsels.

The early Miocene coast held other dangers for pinnipeds. According to John Maisey, an ichthyologist at the American Museum of Natural History, "[t]he diversity of large pelagic sharks has increased since the Late Creta-ceous, with no major declines or extinctions." The modern great white shark's genus, *Carcharodon,* had appeared, perhaps partly in response to a lengthening marine mammal menu. Even full-grown desmostylians might have been fair game for such sharks if they struck from ambush and then waited for victims to bleed to death, as great whites do today. *Enaliarctos* could have fallen victim to white sharks, as the sea otter, *Enhydra,* does now. The sharks don't usually eat sea otters, which lack the fat deposits they crave, but perhaps *Enaliarctos* had more fat, or Miocene white sharks were less choosy.

One thing *Enaliarctos* didn't have to fear, judging from Pyramid Hill and other west coast fossil sites, was reptiles. Although still abundant in the southeast, crocodiles don't seem to have lived on the late Oligocene and early Miocene west coast. This suggests that the Pacific was already signifi-cantly colder than the Atlantic, even though the climate was warm enough to support bald cypress swamps. Currents and upwellings similar to those cooling west coast water today may have contributed to this.

Cool water may have been a factor in enticing *Enaliarctos*'s precursors into the water. Although many tropical marine fish swim up streams today, mass river spawning of lampreys, eels, and salmon is mainly a temperate zone, cool water phenomenon. Lampreys and salmonids probably inhabited the Pacific in the Oligocene, although no fossils from then are known. If they spawned in rivers in anywhere near the numbers they do today (or did, before dams and logging), it would have induced land carnivores to prey on them, as bears do today. If they congregated in bays and estuaries before moving upstream to spawn, it would have induced aquatic carnivores to dive for them, as *Enaliarctos* probably did.

Similarities between the early Miocene and modern coasts have their limits, of course. Baleen whales may have been less diverse when *Enaliarctos* lived than today. The Pyramid Hill area has yielded far fewer baleen species than toothed ones, and Lawrence Barnes thought this might be "a real reflection of the North Pacific assemblage in the early Miocene." There may have been less of the nutrient-rich upwelling that fosters small fish and crustacean prey than today. Although benign, the world of *Enaliarctos, Behemotops,* and the penguinlike *Tonsala* may have had a certain austerity, perhaps a monotony, considering how long it seems to have persisted relatively unchanged.

That equilibrium began to dissolve by about 18 million years ago, however, and somewhat surprisingly so from a modern viewpoint. While the overall trend since the late Eocene epoch has been toward cooling climate, which we, living in an interglacial age, assume to be normal, there have been many reversals along the way. The early middle Miocene epoch in particular seems to have seen significant warming along North America's west coast.

The warming was not a reprise of Eocene "greenhouse" conditions. Tropical forest and coral reefs did not return to the west coast. Instead, a mixed conifer and broadleaf evergreen forest replaced bald cypress and deciduous hardwoods, suggesting that winters were milder, but rainfall decreasing, particularly in summer. The warming did raise water temperatures significantly and cause an increase in embayments and inshore shallows as some polar ice melted and sea level rose. Such changes may have prompted the pinnipeds to begin dispersing from North America's west coast while they allowed a largely tropical marine mammal group to invade—or reinvade—it.

TEN

Sea Cows and Oyster Bears

Climatic warming is complicated, even on the rudimentary level of the seasons. At Point Resistance, midsummer seems paradoxically dank as the rising air of hot inland valleys pulls sea fog over the coast. Despite the fog, the headlands are dry and brown, and after the murres leave, the point looks forlorn, with a few pelicans and cormorants crouched on a rock that seems shrunken. Even harbor seals are scarce. Wind heaves the sea into iron wavelets without the lacy surf of sunny times. Dead kelp clots beaches.

But summer fogs are only a superficial dimming of coastal vitality. Indeed they are part of it, since the cold water upwelling that generates them lifts nutrients from the sea bottom to enrich the food chain. Diatoms and other microscopic algae absorb the nutrients, bloom, and feed vast multitudes of the tiny young fishes, mollusks, worms, and crustaceans that larger creatures eat. When less fertile warmer water spreads north up the coast in El Niño years, the impact on local organisms can be drastic. Murres and other seabirds may stop breeding because cold water food fish like herring disappear as upwelling dwindles. Harbor seals, sea lions, and harbor porpoises also may fail to breed. Tropical species like barracuda, striped dolphins, and green turtles invade with the warmer water.

Something similar seems to have happened on North America's west coast during the early middle Miocene epoch, on a much grander scale. The reasons for the warming remain obscure, but the scale is evident in the fossil record. Cold water animals disappeared, and warm water ones replaced them, although the transition was complicated, involving a welter of strange sea creatures.

The ancient pinniped, *Enaliarctos,* finally died out then, a puzzling example, like ichthyosaurs and toothed birds, of a well-adapted, long-lasting organism that disappeared. The warming must have affected it by changing fish populations, but this seems an incomplete explanation for its demise. The waters north of California would have remained significantly cooler, but *Enaliarctos* disappeared everywhere. Another complication to explaining its extinction is that a number of other pinniped genera appeared as it vanished.

Its earliest known contemporary was *Desmatophoca,* the genus Thomas Condon named in 1906. As Jacob Wortman noted, *Desmatophoca's* back teeth were transitional between land carnivores' sharp-cusped, shearing molars and the simpler "homodont" back teeth of seals and sea lions. *Desmatophoca's* back teeth imply that its diet differed significantly from *Enaliarctos's,* and that its life did too. Other characteristics suggest this.

Desmatophoca was larger than *Enaliarctos,* about sea-lion size, and it was sexually dimorphic, the males significantly bigger than females. It thus probably bred in colonies, unlike *Enaliarctos,* with males competing to dominate harems. Since colonial breeding usually entails a wider-ranging life than solitary pairing, *Desmatophoca* probably swam farther offshore than *Enaliarctos* and may have migrated to areas of seasonal food abundance. While *Enaliarctos* stayed scattered along the coast, chasing good-sized fish in relatively shallow water, *Desmatophoca* may have been many miles out, diving much deeper after schools of small fish, squid, and crustaceans. It had larger eyes than *Enaliarctos,* and its ears were better adapted to directional location of underwater prey.

Desmatophoca's more pelagic life may have given it a survival advantage as conditions warmed. Swimming farther and diving deeper, it may have tapped food sources unaffected by the warming. But other new pinnipeds from that time complicate this. In 1979, Lawrence Barnes described a genus, *Pinnarctidion,* "little fin bear," from slightly higher in the Pyramid Hill strata that produced Emlong's skeleton. It resembled *Enaliarctos* more than *Desmatophoca.* In 1989 and 1992, Barnes described two genera from northern Oregon's Astoria Formation, *Pteronarctus* and *Pacificotaria,* which, although they lived several million years after the Pyramid Hill fauna, resembled *Enaliarctos* even more than *Pinnarctidion.*

If creatures like *Enaliarctos* had survived longer on the coast, then why had *Enaliarctos* disappeared? The three new genera were based on less complete fossils than Emlong's skeleton, so it was harder to tell how they'd lived. And, despite the resemblance, they also had characteristics suggesting they led a more pelagic, colonial life than *Enaliarctos*. So they may have had a survival advantage over the older genus.

In 1995, Barnes described yet another genus from Astoria Formation skulls and postcranial bones Emlong had collected in 1977. Naming it *Proneotherium,* he saw it as possibly the earliest known relative of the odobenids, the group that includes the walrus (*Odobenus rosmarus*). Annalisa Berta and Thomas Deméré, another San Diego paleontologist, saw *Proneotherium* as definitely related to odobenids: "The phylogenetic position of *Proneotherium* as a basal odobenid is confirmed. Cladistic analysis supports monophyly of the Odobenidae, defined here as the clade containing the most recent common ancestor of *Proneotherium, Odobenus,* and all of its descendants."

The walrus, with its storybook whiskers and tusks, certainly seems part of a distinct evolutionary line. It alone resembles *Enaliarctos* in using all four limbs for swimming, and it is otherwise unique. Walruses can rotate their hind legs forward to push them overland, like sea lions, but prefer to hump along on their bellies, like seals. They also resemble phocids in lacking external pinnae. They live in groups, but since they feed mainly on shallow water mollusks, they don't make pelagic migrations to rookeries, although they travel seasonally to follow food availability. Males are larger than females, but instead of fighting to claim harems, they perform elaborate visual and auditory displays in water near where females are pulled out, trying to entice them into aquatic mating.

The recent, highly specialized walrus is not necessarily typical of ancient relatives. *Proneotherium,* the putative earliest known odobenid, was not a specialized clam eater. Like the other post-*Enaliarctos* genera, it had teeth that were evolving toward catching and holding prey in the water rather than cutting it up on land. "This morphologic series reflects a functional change from a shearing dentition to more of a piercing dentition," Berta and Demere wrote, "and from a dentition capable of processing food (i.e. chewing) to a dentition that serves primarily to seize and hold prey." So *Proneotherium* may have been at least partly pelagic and colonial.

The appearance of so many new pinniped species on the west coast in the early mid-Miocene suggests that *Enaliarctos* may have disappeared not just because of warming climate but for the good Darwinian reason that better-adapted contemporaries crowded it out. Still, the fossil record doesn't prove this either. If *Enaliarctos*'s contemporaries occupied different ecological niches from its shoreline-fishing one, they might not have competed with it enough to exclude it.

Perhaps another, yet-undiscovered pinniped genus replaced *Enaliarctos* in the solitary, coast-hugging niche, although none is known. Perhaps some other organism did. The giant plotopterid diving birds common in the Oligocene might seem possible competitors. They had their heyday at about the same time as *Enaliarctos*, however, and also dwindled by the mid-Miocene. Indeed, paleontologists have speculated that the diversification of pinnipeds contributed to the penguinlike plotopterids' decline.

As the pinnipeds bewilderingly diversified and dispersed, a group with a less complicated relationship to warming climate was spreading up the coast. Several genera of dugonglike sirenians had evolved in the Caribbean and Atlantic by the Miocene. Edward Cope named one, *Dioplotherium,* from some teeth found in South Carolina. It was over ten feet long, with two upper incisors prolonged into thick tusks that protruded slightly from a downward-pointing mouth. Another genus, *Metaxytherium,* also had a downturned mouth, but smaller tusks. Another, *Dusisiren,* had a less down-turned mouth and lacked functional tusks.

Daryl Domning, who helped describe Emlong's *Behemotops,* thought the diverse sirenians coexisted because they used tropical sea grasses like *Thalassia,* turtle grass, differently. *Dioplotherium* raked up grass beds with its tusks, feeding on nutritious rhizomes as well as leaves. *Metaxytherium,* a less specialized bottom-feeder, ate the diverse plants that grew where *Dioplotherium* had disturbed the bottom. *Dusisiren* fed on floating plants as well as bottom ones. "The large-tusked dugongs would have acted as key-stone species in the ecosystem," Domning wrote, "keeping both sea-cow and sea-grass diversity at higher levels than they otherwise would have attained."

Fossils recently found in Alta and Baja California show that *Dioplotherium, Metaxytherium,* and *Dusiren* reached the west coast in the early middle Miocene, probably swimming through the Panama seaway and following a northward spread of sea grasses as water warmed. "At that time," Domning wrote, "California had both a tropical climate (surface temperatures perhaps 8 degrees Centigrade warmer than today) and extensive areas of protected shallow waters which must have been an ideal sirenian habitat." They would have been an impressive addition to the fauna. I once looked out a hotel window at Miami airport and saw a manatee, about the size of the fossil genera, filling a golf course canal almost from bank to bank. Three species grazing at all different levels of meandering tidal sloughs would have cut a swathe.

The invading sirenians, of course, encountered the native saltwater plant eaters, the desmostylians. Their fossils often occur together. According to Domning: "The desmostylians probably shared the tropical grasses with [*Dioplotherium*] in the southern latitudes, in addition to harvesting benthic algae and (in more northern parts of their range) cold-water grasses." Because both specialized in bottom-feeding on sea grass, big-tusked *Dioplotherium* may have competed with desmostylians, but if they did, it perhaps was in a sedate way, since they would coexist on California's coast for millions of years. (At least, living sirenians are fairly sedate; desmostylians may have been as rowdy as hippos.)

Many other tropical organisms shifted north as water warmed, and California coastal fish and invertebrate populations changed markedly. Sirenian bones occur with those of warm water seabirds, turtles, and sharks. The toothed baleen whales, *Aetiocetus* and its family, disappeared around this time. Some other primitive groups survived—toothed whales named squalodonts after their molars, which resembled shark's teeth, and toothless mysticetes named cetotheres, "whale beasts." Early versions of some modern cetacean groups appeared, including odontocetes such as beaked whales, sperm whales, and dolphins, as well as mysticetes such as right whales.

New west coast inhabitants were not confined to diversifying pinnipeds and cetaceans or invading southerners. At least one seems to have come from inland. While *Enaliarctos* was trailing off mysteriously a distant relative was starting an even more obscure career. Another aquatically inclined carnivore had lurked in the west coast's Oligocene swamps, and its descendants also went to sea after a fashion. In 1960, R. A. Stirton, an American Museum paleontologist, described a beast he named *Kolponomos* from a partial skull and jaw in a Miocene marine formation on the Olympic Peninsula. He thought it was a large procyonid, a raccoon relative, that had "adapted to live in marine waters." Given raccoons' propensity for frequenting shorelines, this made sense. One living South American species is called the crab-eating raccoon.

Then Douglas Emlong made another discovery near Newport, Oregon— a nearly complete skull and jaw, with some postcranial bones, of an early Miocene creature. The bones were in a concretion, a cementlike sphere of sediment that had formed around them. Emlong first found the back of the skull and some vertebrae and foot bones in 1969. Eight years later he

Figure 13. *Kolponomos* was one of the strangest marine mammals that ever lived, a relative of bears that specialized in diving for mollusks.

located the concretion's other half, with the cranium and jawbone, put them all together, and surmised that they came from Stirton's *Kolponomos*—a particularly striking example of what Clayton Ray called Emlong's "uncanny, unrational" genius for locating and identifying fossils.

"If he could get that fossil in his hand, and try to imagine anything that old, that was the part that fascinated him," his mother said. "He just *had* to know what it was, and he had to know what strata it came from, and about how old it was." But, as his *Kolponomos* discovery suggests, Emlong's finds seem to have arisen as much from day-to-day familiarity with the living coast as knowledge of the prehistoric one, a familiarity as much aesthetic as scientific:

> He saw an otter up there on the Straits of Juan de Fuca . . . a little old otter on the shore, and he got quite close to that before it raised up and looked at him and took to the water. It swam out a little ways, and then turned around and looked at him as curiously as he was looking at the otter. They admired each other a while, and then the otter swam away. Those little things delighted him.

Emlong painted expressionist seascapes, influenced by Van Gogh, and perhaps regarded the littoral bedrock as a sculptor does, sensing life inside and seeking to free it.

Such things can generate a lot of energy, which Emlong needed. "It is no accident that extensive collections had not been made previously from coastal Oregon," wrote Ray. "Much of the best prospecting is on the bedrock floor of the intertidal zone, much of which is buried under many feet of sand through most of the year, or in places for many years running (some promising areas have never been clean in 20 years of experience) and is scoured clean, if at all, only during violent winter storms." Emlong did a lot of his collecting by rushing to newly storm-scoured beaches and cutting out specimens before the tide returned.

"The waves are immensely fascinating," he enthused. "During the winter storms, at high tide huge rollers batter the rocks, sending white spray a hundred feet in the air. . . . Boulders weighing half a ton will be moved 25 or 30 feet, thundering like cannon shots." He went out early to watch the storms. "That was part of the excitement. He even liked to get up in the middle of the night if there was a storm . . . go out and walk in the storm."

Once exposed, the fossils had to be cut from rock hardened by tectonic stress. "[T]he Oregon materials are the most difficult ever encountered by our laboratory," Ray observed. "A combination of techniques has proven essential, no one of which has provided the golden key: old fashioned hammer and chisel, pneumatic hammer, dental burrs and grinder, rock saw, acid digestion, dental and industrial sand blaster." Actually, Ray considered this a good thing in that Emlong's enthusiasm might have destroyed softer specimens. Hammer and chisel were all the equipment he could afford, and he did little to protect fossils once he had them. A load of army surplus plaster bandages driven from Washington, D.C., sat in his yard unused. He packed specimens in newspapers, not bothering to wad them, so that the weight of flat paper in his shipments could exceed that of the fossils.

When I saw Emlong's *Kolponomos* skull at the Smithsonian, it looked as hard as semiprecious stone, and I thought it might be a cast in some superplastic. Even the collections manager, David Bohaska, had a moment's doubt whether it was the real fossil. Its preparation had taken two decades. In the end, however, it showed that *Kolponomos* had been something stranger than a big crab-eating raccoon. Tedford, Barnes, and Ray, who described the skull in 1994, decided it had been an early bear relative rather than a procyonid, but one unlike any other known bear relative. The skull had a downturned snout and particularly broad heavy molars that distinguished it sharply from terrestrial ursids.

"The crushing cheek teeth would have been suited to a diet of hard-shelled marine invertebrates," they wrote. "The anteriorly directed eyes and narrow snout indicate that *Kolponomos* could view objects directly in front of its head." These features, combined with large attachments for neck muscles, suggested that the beast "fed on marine invertebrates living on rocky substrates, prying them off with the incisors and canines, crushing their shells, and extracting the soft parts." The robust foot bones Emlong found also hinted at a mollusk diet, since they could have been used to pry shells loose. In 1999, Annalisa Berta and James Sumich wrote: "*Kolponomos* represents a unique adaptation for marine carnivores, its mode of living and ecological niche are approached only by the sea otter."

The rest of *Kolponomos*'s skeleton remained unknown, and paleontologists were more cautious in making assumptions about it than they had been with desmostylians. The skull lacked the aquatic specializations that linked *Enaliarctos* to the pinnipeds, suggesting that the "oyster bear" swam to mollusk beds with legs relatively little changed from nonmarine ancestors'. "The few postcranial bones available indicate that *Kolponomos* was amphibious but was not a strong swimmer," wrote Tedford, Barnes, and Ray. "Its foot bones clearly indicate retention of significant ability for terrestrial locomotion and an amphibious existence. It was probably littoral in distribution." But only a complete skeleton like Emlong's *Enaliarctos* could prove such inferences. No creature quite like *Kolponomos* lives today or seems to have lived since the Miocene, so it is hard to imagine one. Even its size remains unclear, although Emlong's skull is like a small bear's.

Another thing that remains unclear is the relationship between *Kolponomos* and the early Miocene's warming climate. Warming wouldn't necessarily have increased populations of the littoral invertebrates on which it fed. It might have decreased them if nutrient levels in the water declined. A bear-sized creature that lived by prying shellfish off rocks would have needed a lot of shellfish. Evolution of many new pinnipeds also seems odd in a warming period, since large, diverse pinniped populations mainly live in temperate and arctic waters today. Still, the sirenians' mass arrival proves west coast water was getting warmer. So another factor than climate may have fostered *Kolponomos* and the pinnipeds.

Geology is a possibility. Although the west coast seems to have been low and marshy in the Eocene and Oligocene, like today's gulf coast, it grew increasingly hilly in the Miocene as tectonic activity increased. Volcanoes appeared inland, and offshore waters probably deepened. *Kolponomos* seems to have been adapted to clambering around on rocky headlands and reefs, which would have provided new habitat for mollusks like mussels and oysters. Early mid-Miocene pinnipeds also may have favored rocky shores and reefs, as some of today's west coast pinnipeds do. Healthy sea lions seldom bask on a beach if they can find rocks.

Even as climate and coastal waters were warming, tectonic deepening of offshore waters could have caused increased nutrient upwellings. There are such places in the tropics today. On the Galapagos, California sea lions, fur seals, and penguins live on the equator, nose to nose with marine iguanas. *Desmatophoca* and its pinniped relatives could have lounged on rocky reefs while *Dioplotherium* and its sirenian relatives luxuriated in warm bays—a kind of all-purpose marine mammal paradise. Again, such ideas are speculative. But rising land may well have had something to do with increasing west coast marine biodiversity, because that biodiversity was trending toward its highest level since the Mesozoic.

ELEVEN

The Long, Warm Summer

Living elephant seals seem to have taken the pelagic, sexually dimorphic way of life begun by early pinnipeds like *Desmatophoca* as far as it can go. They breed on beaches a few miles up the coast, but I've never seen one at Point Resistance. They don't feed near shore, so they don't wander along it as seals and sea lions do, instead traveling hundreds of miles and diving thousands of feet to find abundant deepwater prey like squid. Males feed offshore of British Columbia; females in the central Pacific. When not feeding, they frequent large breeding colonies in small areas, navigate to them with precision, and produce a lot of offspring.

Elephant seal bulls are more than twice as large as females. I'll never forget the first beachmaster bull I saw, at Año Nuevo, the biggest rookery on the northern California coast today. He was behind a willow thicket, but his head protruded on one side, his tail on the other. He may have been eighteen feet long, but I didn't try to get a closer look. Competition for females makes bulls extremely combative, although strife seems less evident at the much smaller Point Reyes rookery than at Año Nuevo. Young bulls shift around and spar a lot there, but I've never seen the largest beachmasters inflate their trunklike proboscises and attack anything. In fact, I've never seen the largest bulls at Point Reyes move.

Impressive as elephant seals are, however, they are not unprecedented. Fourteen million years ago, a creature named *Allodesmus,* "other swim-

Figure 14. Although it lived 14 million years ago, *Allodesmus,* "other swimmer," may have been almost as specialized for ocean life as are living elephant seals.

mer," probably resembled them. Males were huge and may also have had inflatable proboscises, which they used to impress females and intimidate other males. And they may have rivaled or even exceeded elephant seals in other respects, swimming great distances to feed, and breeding in enormous colonies. If so, *Allodesmus* might be said to have taken the pelagic way of life as far as it could go, with elephant seals evolutionary copycats.

Allodesmus seems to have reflected its time. If west coast marine tetrapod evolution in general had a peak, a time when it went as far as it could go, the mid-Miocene may have been it. Again, this challenges the tradition that life always tends toward greater adaptation and diversity. But the fossil record suggests that marine tetrapods, at least, were never more diverse and intricately adapted than during a Miocene heyday, even though that time came relatively early in some groups' evolution.

The greatest known west coast assemblage of Cenozoic marine fossils is from that time. Since the 1850s, bone hunters have worked an extraordinary mid-Miocene marine deposit on a hill northeast of Bakersfield, not far from where Emlong found his *Enaliarctos* skeleton. A stratum near the hill's summit, formed between 15 and 13 million years ago when an inland waterway known as the Temblor Sea covered the area, consists of a three-foot layer of bones and other materials that has been described as "the most diverse and best documented fossil marine vertebrate faunule on the North American margin, or indeed from the Pacific Basin." The place is called Sharktooth Hill for its most common fossils, but other kinds may number a hundred per cubic foot.

Sharktooth Hill's reputation probably attracted Emlong to California. As his discoveries showed, more than a century of collecting hadn't exhausted the Bakersfield area's potential. Along with *Enaliarctos,* his 1975 finds in the earlier Pyramid Hill sediments included a "tiny baleen whale," a large toothed whale, and basking shark gill rakers. In 1977, he found the "shell and bones of a gigantic sea turtle" in the area.

Nobody is sure what packed Sharktooth Hill so thickly with fossils. In 1911, F. M. Anderson, a geologist, suggested that "an epoch of violent volcanic activity and fall of ash may have been responsible for the death of large numbers of pelagic animals." Remington Kellogg, who studied the bed in the 1920s, thought Anderson's explanation didn't "satisfactorily account for the scattering and mingling of bones of many species of animals." Kellogg observed that the skeletons would have been complete and articulated if an ashfall had buried them, and that "some agency other than volcanic activity must have been responsible for scattering the remains," perhaps tidal currents.

"The deposit at Sharktooth Hill suggests a minor tidal basin . . . such as one finds today along the eastern shore of the Gulf of California, for example, in the region of Tapolobampo or of Mazatlan," wrote another early observer. "Whales, sea lions, sharks and maritime birds help to sketch in a picture of a fairly shallow embayment into which tidal currents eddied quietly through many years of Miocene time. Drifting in from deeper waters or floating off of nearby sand bars, the carcasses of various vertebrates accumulated here to finally settle to the bottom of relatively quiet waters."

In the 1960s, Edward Mitchell suggested that mass mortality from a toxic bloom of dinoflagellates had caused the accumulation. In 1972, Lawrence Barnes seemed to agree with Kellogg's less dramatic assessment: "I believe that the bonebed may be more easily explained as the result of accumulation of bones on the sea floor over extended periods of time with little or no sedimentation." Barnes estimates that the bones could have piled up over a million years or more.

Kellogg named *Allodesmus,* the Miocene elephant seal look-alike, from Sharktooth Hill fossils in 1922, although rudimentary knowledge of pinniped evolution made interpreting its bones difficult. "Nevertheless," he wrote, "*Allodesmus* is quite distinct from the sea lion . . . and unquestionably represents a more generalized type." In 1960, finding the traditional method of digging into the bed from the sides "unsatisfactory for plotting positions of bones and for gathering other data," Edward Mitchell brought in a D8 bulldozer and scraped a 5,000-foot-square section of the hill down to just above the bone level. This revealed "a nearly complete skeleton of *Allodesmus kelloggii*" that had lain in "relatively quiet water" long enough for a shark to lose a tooth in it and for a whale rib to fall across its backbone. Although broken and displaced, its bones were complete enough for Mitchell to reconstruct the creature.

"*Allodesmus* is a specialized otariid pinniped that diverged from the main line of sea lion evolution," Mitchell wrote in 1966.

> It was a large animal, attaining greater lengths than do living sea lions. It was highly adapted for aquatic life. Its large eyes suggest light-gathering ability at depths, and its heavy ear ossicles indicate adaptation for underwater hearing. . . . The small size of the flippers, the large body size, and other factors indicate that *Allodesmus* did not have a heavy coat of fur. By analogy with phocids, *Allodesmus* probably did not have external pinnae [ears]. Males of *Allodesmus* had a large baculum [penis bone] and a well-developed proboscis, both features characteristic of a high degree of sexual dimorphism.

In other words, *Allodesmus* had resembled the elephant seal overall, the males in particular having been great blubbery beasts with lustful, pugnacious dispositions in rookeries. Unlike elephant seals, they perhaps could push themselves along on land with their back flippers. But if by some chronological twist, an *Allodesmus* bull showed up on a Point Reyes beach today, it might be hard to distinguish from nearby elephant seal bulls.

Allodesmus's behavior is less clear than its anatomy. What and where it ate will remain unknown unless skeletons turn up with prey remains in their abdomens. In some ways, it may not have been as well adapted to pelagic life as elephant seals. Its ears were not as developed for directional hearing as later pinnipeds', and it thus may not have been able to feed as deep. Still, its specialized anatomy suggests that it was able to travel and dive very far to reach food resources unavailable to lesser creatures. Its fossils have turned up in many places on the west coast. Three *Allodesmus* species are known, one of which probably was even more like an elephant seal than Kellogg's original fossil.

Allodesmus was closely related to Thomas Condon's earlier genus, *Desmatophoca*. How the two are related to other pinnipeds remains unclear. Although Jacob Wortman thought *Desmatophoca* was more like a phocid seal than an otariid sea lion or fur seal, most later paleontologists assumed, like Mitchell, that both *Desmatophoca* and *Allodesmus* were "diverged from the main line of sea lion evolution." This seemed to make sense, in that Oligocene phocid fossils were unknown from the Pacific, whereas *Enaliarctos* was considered ancestral only to otariids.

In 1994, however, Annalisa Berta and André Wyss proposed that *Desmatophoca* and *Allodesmus* actually were more closely related to phocids than to otariids. "A number of features are shared among phocids and *Allodesmus* and *Desmatophoca*," Berta wrote, "and hence they support a close link between these taxa." Emlong's work again shaped the phocid-otariid controversy in 2002 when Berta and Thomas Deméré published a paper on numerous *Desmatophoca* bones the collector had amassed in Oregon. The bones provided a better sample than those from which Thomas Condon had described the genus, and Berta and Deméré used them to argue that *Desmatophoca*'s skull morphology and other features were like phocids'.

"Phocids are shown to be the sister group to the desmatophocids," they wrote. "[D]ivergence of desmatophocids from phocids likely occurred sometime before 18 Ma, probably in the North Pacific Ocean basin." And this seemed to make sense if *Enaliarctos* was an early relative of phocids as well as otariids, and if *Desmatophoca* and *Allodesmus* indeed were more like seals than sea lions. Still, despite similarities, both long-extinct genera differed greatly from living animals, either seals or sea lions, so linking them definitively to either group remains problematic.

Sharktooth Hill yielded fossils of other pinnipeds. In 1931, Kellogg named one of these *Neotherium*, "swimming beast," from foot bones found in the bone bed. As the name implies, *Neotherium* is considered a later relative of *Proneotherium*, the possible walrus precursor Emlong found in Oregon. Further *Neotherium* discoveries from elsewhere suggest that it was a relatively small, but sexually dimorphic, animal.

Another Sharktooth Hill pinniped wasn't described until 1988, when Lawrence Barnes named it *Pelagiarctos*, "ocean bear." It was a large relative of *Neotherium* with primitive, meat-shearing cheek teeth like those of *Enaliarctos* but fused lower jawbones convergent with those of living walruses. "Its tooth and mandibular morphology, large body size, and extreme rarity in the fossil assemblage," Barnes wrote, "suggests that it may have been a predator of large marine vertebrates." It seems to have been the pinniped equivalent of a lion or hyena, a top predator with teeth several times larger than those of squid eaters like *Allodesmus.*

Unlike *Allodesmus,* a resurrected *Pelagiarctos* would be easy to distinguish from the seals and sea lions at Point Reyes today—a heavy-jawed brute with fangs instead of tusks. It might have raided its fellow pinnipeds' pullouts and rookeries or perhaps stalked and ambushed them offshore. It is one good indication of a west coast marine tetrapod diversity peak in the mid-Miocene, because such a pinniped megapredator has not occurred here since. Steller's sea lion will eat smaller sea mammals if it can catch them, but isn't equipped to butcher prey, as *Pelagiarctos* was. Of living pinnipeds, only Antarctica's leopard seal, which preys habitually on penguins and smaller seals, might be comparable to it.

Sharktooth Hill's pinniped fossils suggest a very different west coast from the quiet, swampy shore inhabited by *Enaliarctos* and the early desmostylian *Behemotops* 10 million years earlier. If the abundant *Allodesmus* was as contentious as living colonial pinnipeds, the neighborhood of its breeding colonies would have been raucous. If *Neotherium* gravitated to nearby sites, the turmoil would have multiplied, especially if *Pelagiarctos* was on the prowl. And Sharktooth Hill shows that pinnipeds weren't the only big tetrapods paddling along the shore. *Desmostylus* occurred there, and two turtle species, one related to the living green turtle, also a feeder on sea-grass beds, the other belonging to a giant extinct group.

Offshore diversity would have dwarfed beach crowds. Sharktooth Hill fossils show that there were at least seven times as many cetacean as pinniped species in mid-Miocene coastal waters. In 1931, Kellogg called Sharktooth Hill "the largest fauna of cetacea thus far known from any formation on the Pacific Coast of the United States." In 1976, Barnes listed twenty cetacean taxa from the bone bed and estimated that thirty species lived off the west coast in the mid- to late Miocene. "Such diversity," he wrote, "suggests that physical space and potential food resources in the past marine environments were being significantly divided by these various cetaceans." Recent counts show that Sharktooth Hill had more kinds of sperm whales and platanastids (long-beaked toothed whales) than other mid-Miocene assemblages, and that its diversity of kentriodontids (another long-beaked odontocete group) and primitive mysticete cetotheres rivaled other fossil sites.

Fish were equally diverse, and, in some cases, big. Kellogg attributed the broken and scattered condition of the Sharktooth Hill bones partly to one group's activities: "Unusually large numbers of sharks abounded in the Temblor Sea, and pelagic mammals undoubtedly suffered from their predatory proclivities. The skeletons of their prey would be dismembered and their bones scattered." Sharks shed their replaceable teeth while feeding, which explains why they are "by far the most abundant fossil remains found at this locality." Mitchell listed nineteen chondrichthyan genera from Sharktooth Hill, almost all of them extant today: eagle rays, sting rays, angel rays, and skates; smaller sharks like dogfish, hornsharks, brown sharks, and soupfin sharks; bigger ones like sixgills, sevengills, hammerheads, tiger sharks, bull sharks, mako sharks, and more than one species of *Carcharodon,* the great white, which doubtless were eating large marine mammals by then.

Sharktooth Hill also yielded an unusual diversity of marine bird fossils. In 1930, Alexander Wetmore listed a shearwater, a gannet, and a duck relative "strongly suggestive of the modern Canada goose, differing . . . in much larger size" from the bone bed. In 1966, Mitchell listed ten taxa, including an avocet and an albatross as well as shearwaters, gannets, and geese. That same year, Hildegarde Howard, who discovered the penguin-like plotopterids, wrote that Sharktooth Hill had "the greatest number of recorded Avian taxa" in west coast marine beds. And bones of an even more spectacular bird she had discovered turned up there.

Like Emlong, Howard and her southern California colleagues collected fossils by a kind of storm chasing, although they pursued a cultural storm instead of a natural one. As history's biggest urban explosion raged across the landscape, paleontologists hurried about after rumors of fossils exposed by bulldozers and earthmovers. In 1957, at a flagstone quarry near Santa Barbara, Howard found a fossil that could be a mascot for megalopolitan gigantism, the skeleton of a "long-winged, short-legged flying bird" that had soared over the Miocene Pacific. She estimated its wingspan at sixteen feet, but even that was not its most remarkable feature. Its beak bore "bony, tooth-like projections on both upper and lower jaws," and she thus named it *Osteodontornis,* "bone-tooth bird."

Shades of O. C. Marsh's *Hesperornis* and *Ichthyornis*—and a related fossil actually had figured in the ferment over his toothed birds. In 1873, Richard Owen had acquired bones of a huge, snaggle-beaked marine bird from the English Eocene, which he'd named *Odontopteryx,* "tooth wing." Soon after Marsh's Kansas discoveries, the anti-evolutionist Owen may have hoped that *Odontopteryx* would throw a non-Darwinian light on the toothed-bird phenomenon. Examining one of its apparent teeth, he saw that it was "a bony process of the jaw bone," lacking dentine and enamel, not like a real tooth. Perhaps toothed birds were less of a reptile-to-bird link than Darwinians thought.

"When we are favored with the description and figures of the Odontornithes by their accomplished discoverer," Owen wrote, "we shall possess grounds for judging of the ordinal and higher relationships of affinity between the Eocene toothed bird and the Cretaceous *Ichthyornis.*" But, although he thought his *Odontopteryx* resembled living pelecaniformes, Owen never got around to judging those "higher relationships," perhaps because the only other such fossil known before Howard's was a skull acquired by a German sailor in Brazil in 1905.

Owen would have been disappointed, anyway. Howard decided the tooth-beaked giants were different enough from other birds to be placed in a new order, but one that represented no insights into the "higher relationships" of Marsh's *Ichthyornis.* "These birds are primitive only as the order to which they belong may represent an early connection with procellariform-pelecaniform stock. Their 'teeth' are in no wise an indication of primitiveness; rather these bony tooth-like processes represent a development of the jaw bones that followed the loss of true vertebrate teeth and diverged from the basic avian form."

In its way, Howard's *Osteodontornis* was more "evolved" than living birds, none of which have bony toothlike processes on their beaks, although some have slightly toothlike horny projections. Like the pinniped mega-predator *Pelagiarctos,* it seems to have occupied an ecological niche that later became extinct. If so, the mid-Miocene might be seen as a peak of marine birds' evolution as well mammals'.

Osteodontornis used its long, hooked beak in "capturing live fish of large size," probably plucking them from the surface as it cruised. Albatrosses, the biggest living seabirds, catch fish similarly but have maximum eleven-foot wingspans instead of sixteen. And *Osteodontornis* wasn't the biggest of its order. The Brazilian genus, *Pseudodontornis,* "false-toothed bird," attained a twenty-one-foot wingspan. Its bones occur in Oregon.

For all its richness, the Sharktooth Hill fossils must represent only a fraction of mid-Miocene west coast diversity. There would have been other big birds feeding on the huge biomass. The turkey vultures that patrol Point Resistance today and the California condors that once ranged north to British Columbia have fossil ancestors beginning in the Eocene, some with wider wingspans than condors' ten-foot ones. Condors fed largely on marine carcasses, and reintroduced birds, sometimes slow to learn about

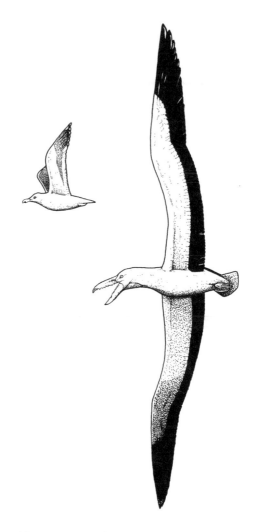

Figure 15. *Osteodontornis,* "bone-tooth bird,"
had a sixteen-foot wingspan and probably
cruised offshore like an albatross, plucking
fish from the surface with its "toothed" bill.

scavenging on land, have taken to beaches on their own. Turkey vultures are
always on carcasses as they appear, and stay on them. When storms buried
a large sea lion near Point Resistance one winter and then exhumed it the
next fall, vultures relocated it before it fell out of the sandbank entombing it.

Extinct California carrion birds called teratorns had wingspans of twelve feet and twice the body mass of living condors. They may have been condor relatives, but, given their greater weight and other traits, ornithologists have doubted this. Storrs Olson thought they might have been pelecaniformes, relatives of the giant Oligocene plotopterids as well as living pelicans and cormorants.

Other pinniped species must have inhabited the mid-Miocene, relicts of early groups like *Pteronarctos,* forerunners of later ones like fur seals and sea lions as well as walruses. There probably were more kinds of desmostylians and cetaceans than have been found. Although sirenians aren't known from Sharktooth Hill, they must have inhabited the Temblor Sea, given their abundance in other mid-Miocene California sites. This was the peak of their west coast diversity, with the three genera that had arrived earlier exploiting all levels of sea-grass beds. If Steller's sea cow behavior was any indication, they would have formed sizable herds at prime locations.

Predators such as *Pelagiarctos,* sharks, and toothed whales would have pushed prey species to find safety in numbers. Throngs of sirenians, desmostylians, and turtles feeding on offshore vegetation while *Allodesmus, Neotherium,* and the "oyster bear," *Kolponomos,* crowded rocks and *Osteodontornis, Pseudodontornis,* and condor ancestors cruised overhead would have been a coastal equivalent of North America's teeming interior savannas, where as many as ten horse species coexisted with proboscideans, rhinos, and other ungulates. Lasting for millions of years, the mid-Miocene must have seemed a kind of endless summer.

TWELVE

Emptying Bays

Summers must end, and the mid-Miocene one began fading about 12 million years ago. Tectonic activity was ongoing, with coastal hills rising higher and offshore waters still deepening and cooling. Antarctic glaciers were growing again and would eventually cover the continent, while the Arctic also would lose the temperate forests that had grown there since the Cretaceous. As though exacting dues for millions of halcyon years, the renewed cooling would bring the greatest mass extinction since the Cretaceous. It hit hardest in the continental interior, decimating the savanna megafauna, but it also affected the coast.

The first marine victims of later Miocene cooling were west coast sirenians. According to Daryl Domning, they had swum into an evolutionary cul-de-sac by invading the California coast 6 million years earlier:

[E]ven when transgressions of warm seas in California created an island of favorable habitat in the North Pacific, their northward spread was inconvenienced by the open coastline of Mexico, which lacked the abundance of protected seagrass meadows that they enjoyed in the Caribbean. Sirenians and, presumably, tropical grasses crossed this barrier and colonized California by the early middle Miocene. The barrier now lay south of them, obstructing genetic exchange with the Caribbean: on the north, their spread was limited by the extent of tropical conditions and/or shallow embayments. . . . Thus California, as sirenian habitat, was a biogeographic island.

Island populations are especially prone to extinction, as the historical disappearance of many has shown. *Dioplotherium* and *Metaxytherium,* the tusked sirenians that fed mainly on bottom plants, may have competed for food with desmostylians during the long mid-Miocene summer, but there evidently had been enough sea grass to go around. As winters cooled, sea level fell, and bays and inland seas drained, the bottom-feeding sirenians were at a disadvantage. Desmostylians were more tolerant of cool water, as their fossil presence around the North Pacific Basin shows, so they would have exploited the shrinking sea-grass supply more effectively.

The *Dioplotherium* and *Metaxytherium* species that had been evolving off California for 6 million years might have survived by withdrawing into warmer waters. As Domning observed, however, the shoreline south of present-day Baja California was mostly straight and deep, without large bays—poor sea-grass and dugong habitat. So, as what had been a sirenian island paradise dissolved around them, the two species dwindled. *Dioplotherium* and *Metaxytherium* haven't been found in California formations less than about 12 million years old.

Sirenians can be surprisingly resilient, however. The modern ones have outlived much more formidable herbivores, and one species survived on California's harshening later Miocene shoreline. *Dusisiren,* the tuskless genus that had invaded with *Dioplotherium* and *Metaxytherium,* could find food even as tropical sea-grass beds dwindled because it also fed on floating and emergent algae, which was increasing as water cooled.

Indeed, the "forests" of holdfast-anchored bull kelp and other large algae now characteristic of the west coast probably did not grow off California before the late Miocene, because the water was too warm and shallow. The huge plants need the abundant nutrients supplied by deepwater upwellings. They presumably had been evolving farther north, although there's no direct evidence of this, since kelp fossils are almost unknown. But there is fossil evidence that *Dusisiren* was exploiting a rich new food source—it began to get bigger. This would have helped it to adapt to cooling water, since heat loss decreases geometrically with increased mass.

Dusisiren adapted so well to a changing North Pacific, in fact, that it eventually outlasted desmostylians. Although distributed along the shores from Baja to Japan, the strange four-legged swimmers survived only a few million years more than *Dioplotherium* and *Metaxytherium.* Cooling, deepening water probably played a part in their disappearance. The sea grasses they depended on need clear, shallow water, since, unlike kelp, grasses don't absorb nutrients through their leaves (a property that allows kelp to grow faster than marine angiosperms and helps to explain why angiosperms don't dominate the shore).

Sea-grass scarcity alone may not explain the extinction of such a long-lasting group. Even today, there is a lot of cool-water sea grass like eelgrass along the Pacific shore, and desmostylians in northern waters must have eaten it. Still, no other explanation is evident. No other known large mammal occupied the bottom-grazing niche after the sirenians disappeared, so competition apparently wasn't involved. Unless deteriorating habitat wiped them out, desmostylians join *Enaliarctos* and its relatives in the lengthy roster of once-prevalent groups whose disappearance is not well understood.

Competition may have affected the mid-Miocene's other dominant marine beast, elephant seal-like *Allodesmus,* which disappeared about 12 million years ago. Cooling, deepening waters may have favored a rival group of pinnipeds. The early walrus relatives, *Proneotherium* and *Neotherium,* were relatively small animals, but the odobenids soon grew bigger. In the 1960s, Edward Mitchell described an impressive one named *Imagotaria* from the late mid-Miocene.

Charles Repenning wrote in 1976 that *Imagotaria* was "sexually dimorphic to an extreme degree" and "was remarkably abundant along the west coast of North America from 12 to possibly 9 million years ago." About the size of a living walrus, it had "homodont dentition similar to modern sea lions, and had specializations . . . that, while being clearly odobenid, imply that it was a deep-diving pelagic feeder with well-developed rookery-type behavior." Repenning cited a locality near Santa Barbara with such abundant *Imagotaria* fossils that it might have been a rookery. Given limited rookery sites, perhaps dwindling ones as the coast grew steeper, *Imagotaria* might have gradually replaced *Allodesmus* in breeding colonies, particularly if adaptations such as improved directional hearing allowed it to feed at greater depths.

Imagotaria evidently ate fish and squid, unlike the living walrus, which feeds by sucking mollusks from their shells. Other odobenids had begun to eat shelled mollusks, however. Lawrence Barnes described a "nearly complete skeleton" of a late Miocene one named *Gomphotaria pugnax,* "pugnacious bolt-toothed otariid," from southern California. He likened its "large upper canine tusks" to those of walruses but found it a different beast otherwise:

> Both upper and lower canines are enlarged and procumbent and worn anteriorly, indicating that the animal may have probed the substrate in

search of benthic invertebrates for food. Extreme breakage and subsequent wear of large, single-rooted cheek teeth indicate that at least some, if not all, the food species (e.g. mollusks) probably had hard shells. Absence of a highly vaulted palate, present in walruses, indicates that *G. pugnax* did not suck bivalve tissues using the tongue-piston method employed by walruses.

Kolponomos, the oyster bear, was another victim of Miocene extinction. And although the first known *Gomphotaria* lived millions of years after the last known *Kolponomos,* pinniped radiation into the mollusk-eating niche might have played a part in its disappearance. Fully aquatic, colony-breeding pinnipeds could have exploited shellfish beds more efficiently than solitary, partly terrestrial beasts. At least, competition seems a more likely explanation for *Kolponomos's* extinction than cooling climate and a steepening shore, neither of which would seem inimical to a bearlike animal.

With fewer sirenian species and no desmostylian herds or oyster bears, the coast of 10 million years ago would have looked emptier than that of 15 million years ago. *Pelagiarctos,* the pinniped superpredator also seems to have disappeared, perhaps because of its large animal prey's decline. Big *Allodesmus* was long gone, and although the walrus-sized *Imagotaria* had replaced it in the pelagic, colonial niche, it and mollusk-eating *Gomphotaria* may have been the only large pinnipeds at that time.

The contrast is so marked that it seems hard to explain just through climatic and geological change. The weather did not get *that* much colder in the late Miocene. There were no ice sheets outside Antarctica, and although warm-temperate savannas disappeared from most of the North American interior, coastal vegetation remained a rich mixture of woodland and grassland. The shore was hillier, but not forbiddingly mountainous. It is tempting, as with Cretaceous marine reptile disappearances, to look for other factors to explain the relatively abrupt and simultaneous extinction of so many large animals. But the late Miocene lacks evidence of major asteroid impacts or other global catastrophes.

As in the late Cretaceous, moreover, many organisms passed through the late Miocene extinction unscathed. Although some early cetaceans like the squalodont toothed whales and cetothere baleen whales dwindled, others remained diverse, and new forms appeared. These included the earliest known humpback whale, *Megaptera miocaena,* identified from a skull found near Lompoc, and the earliest version of living porpoises, named *Loxilithax stocktoni.* Indeed, cetacean diversity may have increased on the

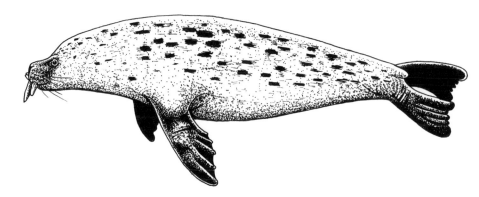

Figure 16. *Gomphotaria pugnax,* "pugnacious bolt-toothed pinniped," was an early relative of the walrus, with tusks in its lower as well as its upper jaw. Like the walrus, it probably ate mollusks.

late Miocene west coast. It is highest in temperate zones today, so cooling water may have allowed more species to frequent the coast.

Fossils suggest that seabirds also diversifed in the late Miocene. Along with the preexistent shearwaters, boobies, auklets, geese, and tooth-beaked osteodontorns, a marine deposit studied by Hildegard Howard at Laguna Hills in Orange County contained bone fragments that may be the earliest fossils of the murres, guillemots, and loons that frequent Point Resistance today. A later Miocene deposit at Oceanside near San Diego contained definite species of murre, guillemot, and loon as well as an albatross and an auklet.

Both formations also contained a new flightless seabird, an auk named *Praemancalla.* (Auks and auklets are, respectively, larger and smaller relatives of murres.) It is interesting that flightlessness seems to have evolved as coastal water was cooling in the later Miocene, since the flightless plotopterids had lived when water was cooler in the Oligocene. Local seabird food supplies may have grown more abundant and reliable as offshore currents became stronger, obviating flight incentives for at least one genus. Tectonic activity might have fostered flightlessness by providing plenty of offshore rocks for *Praemancalla* to nest on safely.

Late Miocene cooling may have produced another reversion to an Oligocene-style fauna. In 11-million-year-old formations, including the one near Lompoc that produced a humpback whale, a new pinniped genus named *Pithanotaria* appears. It was the first known forerunner of living sea lions and fur seals. "This form is recognizable as an otariid largely because of the character of its postcranial skeletal elements," wrote Charles Repenning in 1976,

> although some of these . . . are so primitive that they might not be recognized if they had not been found in association with other skeletal elements. Nevertheless, only size and minor differences in form separate most of the postcranial bones of this earliest otariid from those of living otariids. In describing the holotype of *Pithanotaria starri,* a contemporary of *Imagotaria downsi,* Kellogg (1925) compared it quite favorably with the living Alaskan fur seal.

Pithanotaria was like sea lions and fur seals in having "homodont" molars, suggesting that it caught and ate fish in the water instead of hauling them on land, as Oligocene *Enaliarctos* may have. Yet *Pithanotaria* was unlike its living relatives in other ways. It was small, even smaller than *Enaliarctos,* and, unlike most other known fossil pinnipeds after *Enaliarctos,* it was not sexually dimorphic. "All later otariids have very distinct supraorbital processes in adult males but no specimen of *Pithanotaria* demonstrates the presence of such processes, again suggesting that the genus was characterized by little sexual dimorphism," wrote Repenning. "It is here suggested that the earliest otariid may have been a coastal feeder and that the family had not yet evolved pelagic-rookery behavior."

Almost 10 million years after solitary, coast-hugging *Enaliarctos* had disappeared from the west coast, then, another solitary, coast-hugging pinniped appeared. As with the reappearance of flightless seabirds, this suggests that the late Miocene's cooling waters were reproducing some environmental factor that had existed earlier, and that had encouraged the reappearance of solitary, coast-hugging pinnipeds.

One such factor might be anadromous fish, more typical of temperate rivers than tropical ones. Large numbers of fish entering streams to spawn could add significantly to food supplies for coast-hugging tetrapods. As I've said, lampreys and salmonids may have coexisted with *Enaliarctos* and the penguinlike plotopterids, although Oligocene fossils of them are unknown on the west coast. But, then, anadromous fish fossils are generally rare and obscure. For a long time, the only early west coast salmonid fossil was one that Edward Cope described from Idaho in 1870. He named it *Rhabdofario,* which sounds more like a game of chance than a fish (Cope wrote home from one trip that, having grown a goatee, he looked like "a gambler, or

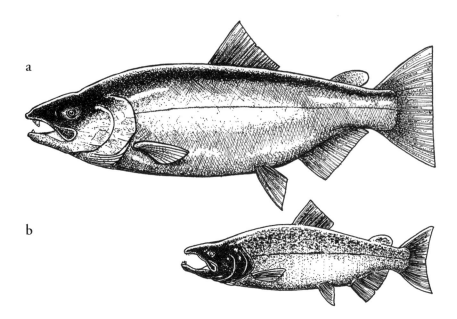

a

b

Figure 17a. *Smilodonichthys,* "saber-toothed fish" (recently reclassified as *Oncorhynchus*), was more than twice the size of the largest living Pacific salmon species. Fossils show that it spawned in west coast rivers.
Figure 17b. Record king salmon.

some other piratical craft"), and although its jaws and snout suggest it was related to living salmon and highly predaceous, little else is known about it. In 1972, two ichthyologists complained that although western salmonids include upward of thirty species, more than half of them endemic, "an astonishingly few fossil specimens have been discovered which can help document the evolution that has taken place in this group."

As it happened, however, the ichthyologists were describing just such fossil specimens. "As early as 1917," wrote Ted Cavender and Robert Miller,

> vertebrae, teeth and skull fragments of an extremely large but unfamiliar type of fish were unearthed, along with mammalian remains, at Pinole, Contra Costa County, California. Only within recent years, however, has it been possible to identify these with certainty as being the remains of an extinct form of salmonid that once was distributed in the coastal regions of the Pacific Northwest probably much the way that Pacific salmon, *Oncorhynchus,* are today. In 1950 and again in 1964, more complete speci-

mens were discovered of this unusual species. . . . The last find consisted of a large skull which is outstanding in its completeness and detail of preservation.

The specimens were mainly from post-Miocene deposits, but the fish was so strange and specialized as to suggest that it had been evolving in west coast rivers for a long time. Cavender and Miller named it *Smilodonichthys,* "saber-toothed fish," because, like the saber-tooth cat, *Smilodon,* it had two upper teeth so large that they protruded below its jaws. *Smilodonichthys* did not use its teeth to catch prey, however, but in mating displays, as living *Oncorhynchus* males do. In fact, *Smilodonichthys* did not have many teeth except the male's sabers. Instead, the structure of its gills suggested that they were "modified for capturing small organisms in a pelagic environment, probably the ocean." *Smilodonichthys* evidently grew gigantic by scooping up plankton offshore, then swam up rivers to breed. Its fossils often occur in freshwater deposits with equid, antilocaprid, canid, and rodent bones. It must have produced huge numbers of offspring, which presumably returned downstream in handy bite-size for shore-huggers like *Pithanotaria* and *Praemancalla.*

Smilodonichthys fossils have continued to turn up, including scales and vertebrae from the late Miocene and early Pliocene Drake's Bay Formation, which crops out in pale cliffs between Point Resistance and Point Reyes. There is also fossil evidence that the living genus *Oncorhynchus,* which includes chinook and coho salmon as well as steelhead and other species, existed in the late Miocene (in fact, taxonomists have reclassified *Smilodonichthys* as *Oncorhynchus*), and little reason to suppose that other groups like lampreys were not breeding in west coast rivers. So late Miocene coastal waters must have been crowded in season, albeit less diversely than the mid-Miocene's.

Dusisiren would have fed on kelp rafts and drunk at stream mouths, while *Gomphotaria* probed the bottom for mollusks and echinoderms. *Imagotaria* perhaps bred on protected land beaches as well offshore islands. Bird colonies must have been large and noisy, and, with murres and guillemots present, they may have resembled today's. Four-foot-long *Pithanotaria* perhaps playfully chased saber-toothed salmon as big as it was, although, given its nonshearing molars, it probably couldn't eat them. Something doubtless ate full-grown *Smilodonichthys,* however, which suggests there is more to learn.

THIRTEEN

Punctuated Pinnipeds
and Darwinian Sirenians

The Miocene epoch was a perplexing evolutionary time as well as a rich one. The apparent absence of a coast-hugging "unisex" pinniped like *Enaliarctos* and *Pithanotaria* from the west coast fossil record between 18 and 11 million years ago is an example. Of course, fossil absence doesn't prove coast-hugging pinnipeds were absent during those 7 million years. *Enaliarctos* may have lived on for millions of years after its apparent extinction, hugging the coast while it evolved into something else. It may have evolved into *Pithanotaria*, although, given the effulgent branchings of pinniped evolution, that would be too simple.

The time gap between *Enaliarctos* and *Pithanotaria* again brings up the issue that has tormented evolutionists since Darwin, the "imperfection of the fossil record." The issue becomes particularly crucial for times of environmental change and mass extinction like the later Miocene, because that is when imperfections become most glaring, as tusked sirenians and desmostylians seem to vanish abruptly from the coast while little otariids and saber-toothed salmon suddenly appear. Since Darwin, most evolutionists have continued to believe that the fossil record gradually will gain continuity as more transitional organisms like Marsh's toothed birds turn up.

When Cavender and Miller were describing saber-toothed salmon, however, two paleontologists, Stephen Jay Gould and Niles Eldredge, began asserting that the fossil record's abrupt appearances and disappearances were not mere imperfections, but evidence of the way evolution works. Authorities, respectively, on fossil snails and trilobites, Gould and Eldredge did not see in them the gradual evolution that Darwin's natural selection predicted. Instead, the invertebrates seemed to remain the same over long periods, then to shift suddenly at times of environmental change. They attributed the sudden shifts not to old organisms evolving into the new ones *in situ,* but to extinction of the old organisms because of environmental change, followed by immigration of new organisms that had evolved elsewhere.

According to the Gould and Eldredge model, "punctuated equilibria," there need be no mystery about *Enaliarctos*'s disappearance 18 million years ago and *Pithanotaria*'s appearance 11 million years ago. *Enaliarctos,* which had stayed largely the same since its appearance 25 million years ago, would have become extinct as the temperate environment to which it was adapted became warmer. In fact, coast-hugging pinnipeds are rare in warm climates today. Then, 7 million years later, when the west coast shifted back to cooler conditions, *Pithanotaria,* which had been evolving where those conditions already prevailed, perhaps off Alaska, would have spread south and occupied a renewed version of *Enaliarctos*'s coast-hugging niche.

Punctuated equilibria has logic. Given the lack of known transitional fossils bridging the 7-million-year gap, it does seem more likely that *Enaliarctos* simply died out, to be later replaced by an only distantly related *Pithanotaria,* than that *Enaliarctos* gradually evolved into something else, perhaps *Pithanotaria.* Still, as with the west coast's scarcity of Eocene cetacean and sirenian fossils, absence of evidence is not evidence of absence. Punctuated equilibria can't guarantee that fossils transitional between *Enaliarctos* and *Pithanotaria* won't unexpectedly turn up. And the late Miocene fossil record may show one classic Darwinian transition between different genera *in situ.*

A few years after Gould and Eldredge announced punctuated equilibria, Daryl Domning maintained that a greatly improved North Pacific fossil record of the tuskless sirenian, *Dusisiren,* showed such a transition. According to Domning, the original invader from the Caribbean, named *Dusisiren reinharti,* evolved "without phyletic branching" into a new species, *D. jordani,* by about 10 million years ago. "*Dusisiren jordani,* under the pressure of recurrent Late Miocene cold episodes . . . proceeded to refine its specialization for open-coast kelp eating, as can be seen not only from its increased body size and the reduced size of its teeth (reflecting an increasing proportion of algae in its diet) but also from incipient neck modifications compensating for its difficulty in maneuvering and feeding in higher energy environments."

Far from disappearing as conditions changed, then, *Dusisiren* gradually became a new kind of sirenian, one able to thrive in the "high energy environment" of a rocky, surf-swept coast with deep, cold water. This was a new environment for sirenians. Although they lived at higher latitudes in the Atlantic during the warm mid-Miocene, there's no evidence that they outlasted the late Miocene chill there. Neither manatees nor dugongs tolerate cold water today.

About 8 million years ago, a new sirenian genus appeared on the California coast, *Hydrodamalis,* whose distant "sea cow" descendants Georg Wilhelm Steller would discover at Bering Island. As Steller would observe, *Hydrodamalis* was different from all other sirenians, over twice as large, with forelimbs more like scrub brushes than flippers, and a mouth containing bony plates for crushing kelp instead of teeth for chewing sea grasses. According to Domning, however, bizarre *Hydrodamalis* was still part of the *in situ* lineage that had begun with *Dusisiren reinharti* 10 million years earlier:

> This evolution is not more than further elaboration of the late Miocene adaptation to algae-eating. Increased access to benthic algae following extinction of the desmostylians may have been important in allowing the growth of *Hydrodamalis* to gigantic size. The colder waters demanded greater body size, and blubber for thermal efficiency; this promoted facultative if not obligate buoyancy, an advantage in feeding near the surface, but also decreased agility and demanded greater care than ever in maneuvering in rocky shallows where they increasingly had to seek their food. Thus was *Hydrodamalis* compelled, unlike any other marine mammal, to compromise an aquatic heritage by a curious venture in the direction of terrestrial life—reduced behavioral or anatomical aptitude for diving and modification of the forelimbs to "walk" through the shallows.

Like the sea creatures in my dream, it seems, *Hydrodamalis,* was performing the unusual feat of returning at least partly from the usually one-way trip of marine adaptation. Or, if a 20-foot, slug-shaped mammal with only tiny forelimbs seems an unlikely returnee to land, then at least *Hydrodamalis* managed to move from the subtropical bay habitat of its *Dusisiren* predecessors on the California coast 15 million years ago to the rocky surf habitat of the same coast 8 million years ago. And it apparently did so by evolving gradually from *Dusisiren* in classic Darwinian fashion, not by branching off in some isolated hinterland and then reinvading when conditions changed.

"Within the California-Baja California province," Domning wrote,

it is not credible that effective barriers to gene flow existed. The embayments were not widely separated and several of the sirenian species are
known from more than one depositional basin. If isolated subpopulations
existed at all, they must have been on the northern periphery, in embayments in northern California, Oregon, or Washington isolated by
stretches of exposed coastline like that of the "Mexican barrier." However,
no trace of sirenians has been found among the many Late Tertiary
marine mammals (including desmostylians) collected from these areas.

Even Douglas Emlong never found sirenian fossils on the Northwest
coast, a good index of scarcity. "Thus," Domning concluded, "the simplest
conclusion is that the hydrodamaline [sea cow] lineage comprised a single
evolving panmictic [all related] population. The punctuated equilibria
hypothesis is unnecessary and, indeed, difficult to defend on the basis of
available data." Domning added in 1978 that *Hydrodamalis*'s new adaptations and tolerance of cold water eventually had allowed it to extend sirenians' range from California around the Pacific Rim to Japan.

A sirenian finally turned up in Oregon in the 1980s, but it did not challenge Domning's *Dusisiren*-to-*Hydrodamalis* transition model. Of unidenti-
fiable genus, the fossil was from the early Miocene, the oldest definitely
identified sirenian from the west coast, and thus had lived long before the
transition, when, as Domning and Ray wrote, "the coastal waters of Oregon
were near the highest temperature they reached during the Neogene [Mio-
cene to present], and still warming." Sirenians might have been predicted
to range north of California at that time, indeed, "it would not be surprising to find that sirenians had reached the eastern North Pacific in the
Oligocene or even earlier."

A 9-million-year-old *Dusisiren* fossil that turned up in Japan in the 1980s
did challenge Domning's model. Its occurrence thousands of miles from
known California *Dusisiren* fossils raised that possibility that, in punctuated
equilibria style, *Dusisiren* might have spread to Japan, where an isolated
population evolved into *Hydrodamalis,* which then spread back eastward,
eventually to replace a less cold-adapted *Dusisiren jordani* in California.

The Japanese *Dusisiren* didn't prove that *Hydrodamalis* had evolved in
Asian waters, however, so it didn't necessarily uphold a punctuationist model.
Hydrodamalis still might have evolved off North America, as Domning
postulated, then spread westward and replaced an isolated Japanese *Dusisiren.*

Or *Dusisiren* simply might have been a more wide-ranging genus than previously thought as it evolved "without phyletic branching" into *Hydrodamalis*. A textbook description of the Japanese species, *Dusisiren dewana*, as "a good structural intermediate between *Dusisiren jordani* and Steller's sea cow in showing a reduction of tooth and finger bones" might support either a punctuationist model or a gradualist one.

A problem with the imperfect fossil record is that even as it grows it doesn't necessarily get more perfect. As with the Japanese *Dusisiren*, a new fossil in one place can call in question an idea carefully constructed from dozens of fossils in another. If natural selection has been called the "law of higgledy-piggledy," then paleontology might be called the science of it, since so much depends on luck. The early Miocene Oregon sirenian became known to science because Lawrence Barnes happened to see it in the house of a man who'd picked it up on the beach.

A related problem with periods of great evolutionary change like the late Miocene is that they can last a very long time. After tusked sirenians and desmostylians disappeared 10 million years ago, there was still another 5 million years of late Miocene epoch to go, plenty of time for more organisms to appear or disappear with apparent abruptness, and many did.

Early walrus relatives developed a particular propensity for entrances and exits in that time. The first large one, *Imagotaria*, having possibly pushed the elephant seal-like *Allodesmus* from the pelagic niche, was the dominant pinniped for 3 million years. It apparently had vanished by 8 million years ago, however, succeeded by the most confusing assemblage of pinnipeds yet. "Between 8 and 4 million years ago no less than seven odobenine genera are known from the North Pacific," observed Charles Repenning in 1976. "They represent a variety of adaptive types; some remained active pelagic predators such as *Imagotaria* and others appear to have progressively adapted to a shallow-water molluscan diet as seen in modern walrus."

Repenning's list didn't even include *Gomphotaria pugnax,* which replaced bearlike *Kolponomos* in the niche of prying mollusks off rocks, since Barnes didn't describe it until 1991. *Gomphotaria* is classed in a group called the dusignathids, which had enlarged tusklike canines on both their upper and lower jaws. Some may have been pelagic fish eaters, but most, like *Gomphotaria,* apparently got a living by cracking mollusks with their molars. Another group, odobenines, had tusks only in their upper jaws, like the living walrus. All odobenines seem to have been mollusk eaters, and some apparently sucked bivalves out of their shells, as living walruses do.

By moving from a pelagic to a more coastal habitat, as Repenning noted, the late Miocene walrus relatives seem to have been undergoing a development similar to that of *Dusisiren* and *Hydrodamalis,* "coming ashore," in a sense. Like the sirenians, they may have done so because their food supply was changing as the coast cooled and steepened. Perhaps mollusks were more abundant in the giant kelp forests that were replacing sea-grass beds. Or clam beds may have increased as cooling waters grew more fertile.

As large, pelagic *Imagotaria* disappeared, large, pelagic relatives of the little coastal otariid *Pithanotaria* seem to have replaced it. "About 8 million years ago," wrote Repenning,

> there were otariids in the North Pacific that definitely showed an increase in body size, had flipper bones that, while still primitive in some respects, were unmistakably otariid, showed massive supraorbital processes on the skulls of the males, and were clearly sexually dimorphic to a degree equal to living otariids. Except for slight differences in some of the limb elements and for the retention of double-rooted cheek teeth, these forms could easily be taken for modern sea lions.

In 1977, the skull of such an animal turned up in the Drake's Bay Formation a few miles north of Point Resistance. It is named *Thalassoleon,* "sea lion," although its bones resembled a giant fur seal's more than a sea lion's. It probably had a dense pelt like a fur seal, although the bones don't show this. It lived along the California coast from about 6 to 4 million years ago. These early otariids did not diverge into as many genera as the walrus relatives, but they established themselves firmly in a colonial, pelagic niche that has lasted ever since, perhaps pushing the odobenids toward a more coastal, mollusk-eating niche in the process. They also became numerous enough by 5 million years ago to move into the Southern Hemisphere, where they spread around the world and evolved into other living genera.

Odobenids never seem to have crossed the equator, and, despite their explosive diversity, most genera did not last long by geological standards. Only a few of the double-tusked dusignathid group outlived the Miocene by much. *Gomphotaria* left no fossils less than 5 million years old. A genus that lived longer, *Pliopedia,* may have been confined to an inland sea, since its bones have been found only in hills around California's Central Valley.

The single-tusked odobenine group also dwindled on the west coast, although ancestors of today's walrus survived. Sometime between 8 and 5 million years ago, they may have swum through the Central American seaway into the Atlantic, where descendants eventually evolved into the living

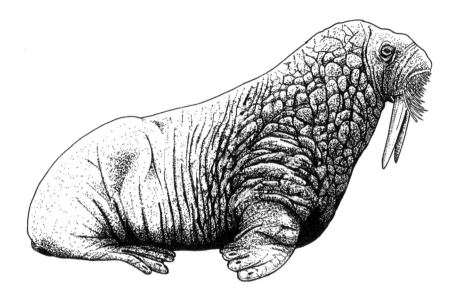

Figure 18. The walrus, *Odobenus rosmarus,* is the last living member of a pinniped group that dominated the west coast for most of the past 20 million years.

Odobenus. Or, as recent fossils from Japan suggest, walrus ancestors may have survived in the North Pacific, later colonizing the Atlantic via the Arctic Ocean.

Within 5 million years, then, walrus relatives branched into perhaps a dozen species and then mostly vanished, making that group the west coast's most conspicuous example of pinniped radiation and decline. It also is the most puzzling example, because neither climate change nor competition seems to explain it, at least in the mollusk-eating niche. The living walrus is a voracious exploiter of clams and other bottom-dwelling mollusks, which it hunts not by digging with its tusks, as was first thought, but by stirring sediments with its snout and whiskers. (Tusks serve for fighting, sexual display, and locomotion—*Odobenus rosmarus* means "tusk-walking sea horse.") It digs up buried clams by squirting water on them, then sucks them from their shells, and can eat six or seven clams a minute, up to 6,000 in a single feeding period.

Walruses are so specialized for clam eating (although they sometimes prey on seals and birds, spearing them with their tusks and sucking out their innards with their vacuum-pump mouths) that their efficiency may limit their success. Today the Pacific subspecies occurs only north of the Aleutians in the relatively shallow waters where enormous clam beds exist. The coast farther south may simply be too deep for it, although that doesn't explain the disappearance of its otherwise adapted relatives. The odobenids' geologically dizzying rise and fall seems a fitting coda to the Pacific Miocene, which might be described in Dickensian style as "the best of epochs, the worst of epochs."

FOURTEEN

Advent of Autumn

It's unfair, really, to call paleontology the "science of higgledy-piggledy." It is no more prone to the unexpected than other life sciences, and, anyway, that is what I like about them. Evolution itself is as full of surprises as fossil hunting. What, for example, might have been predicted to replace ursine *Kolponomos* and tusked pinnipeds like *Gomphotaria* in the mollusk-eating niche? Another massive, tusked animal. But, if anything replaced bearlike and walruslike mollusk eaters on the west coast, it was a different kind of creature entirely, a smaller one that used clever paws and tools to open shellfish instead of piston tongues.

The walrus relatives' late Miocene decline may have given the sea otter an opportunity. Charles Repenning observed that "the oldest marine occurrence" of otters in the eastern North Pacific is in the inland sea beds that contain the odobenid, *Pliopedia,* and that those otters were "showing progressive development toward the living molluscivorous sea otters of the present, in the same area," in beds containing another odobenid genus, *Valenictus.* "It may be that the final 10 million years of odobenid history in the North Pacific resulted in another fissiped [pawed] carnivore entering the sea, enaliarctos-style, and occupying the shallow-water molluscivorous niche left vacant by the dusignathine odobenids."

The otter of California's late Miocene inland sea may not have been the living one's ancestor. It was an extinct genus, *Enhydritherium,* and although it was the size of today's sea otter, *Enhydra* (which is commonly four feet long and weighs fifty pounds or more), it probably swam differently, frequented lakes and rivers as well as shores, and spent more time on land. Late Miocene *Enhydritherium* fossils also occur in Europe and Florida, suggesting that it reached the west coast from elsewhere. Still, as Repenning observed, it did show development toward mollusk eating. Thick, worn premolars on *Enhydritherium* skulls imply that it ate hard foods. And west coast *Enhydritherium* fossils have been found only in marine deposits, suggesting it preferred salt water in this region.

Perhaps *Enhydritherium* began to float on kelp rafts and break shells with stones on its chest, as living sea otters do, although there is no evidence of this. But then, surprisingly little is known about *Enhydra*'s habit of using stones as tools. Early naturalists like Georg Steller did not mention it, and modern ones haven't studied it as intensively as they've studied cetacean behavior. It is so common in California—otters can be seen hammering shellfish on chest stones in every cove along Route 1 south of Monterey—that we tend to take it for granted. But no other living marine mammal or member of the otter family does it.

Sea otters lack opposable thumbs, and their forepaws are tightly webbed, but their manual dexterity is extraordinary. They use rocks not only as chest-held anvils for breaking shellfish, but as hammers for knocking abalones and other sessile animals off substrates. They may have favorite rocks that they carry around in a pouch of loose skin under their forepaw. They may use other objects such as driftwood, shells, and empty bottles to pry prey from the bottom. How the practice evolved is anyone's guess, although other otter species like to play with stones, which probably was a behavioral starting point. It also is unclear how the practice is passed along, although babies probably learn it from their mothers and other otters.

Tool use evidently was a response to the west coast's highly productive giant kelp forests, with their great biodiversity, which offered an opportunity to exploit mollusks and other animals in a more manipulative fashion than *Kolponomos* or *Gomphotaria* had. It's as though *Enhydra* evolved in a marine version of tropical forest, with a similar wide array of trophic niches for clever paws to reach into. Like pet monkeys, tame baby sea otters are adept at picking pockets. Like chimpanzees, wild otters use tools opportunistically, getting food in many other ways as well. They recognize aluminum cans as favored octopus hiding places and bite the cans open. They hunt clams by digging trenches that may be a foot deep and four feet long. And while it took primates perhaps 40 million years to produce a tool-using tropical forest species like the chimpanzee, otters seem to have required only about 5 million once they moved into the kelp forest.

Odobenid decline and the sea otter's unprecedented appearance seem typical of the Pliocene epoch, which followed the Miocene. It was a time of unpredictable change. For one thing, climate warmed again near its beginning. It didn't warm as much as it had in the mid-Miocene, and that epoch's mass immigration of tropical animals did not recur. *Enhydritherium* may have arrived from the Caribbean through the Central American seaway, however. And other creatures may have wandered into west coast waters.

Although an absence of early phocid seal fossils in the Pacific had long supported the idea that they evolved in the Atlantic from a different ancestor than other pinnipeds, this changed in the 1980s. Definite Miocene phocid fossils turned up in Peruvian deposits, showing that they had undergone part of their early evolution in the Pacific. Phocids may have swum into the North Pacific fairly early too, although there is less clarity about this.

In the 1990s, I read about an apparent fossil phocid that S. David Webb, a paleontologist at the Florida Museum of Natural History in Gainesville, had found in the Pliocene Imperial Formation of southern California's Salton Sea Basin. It seemed related to monk seals, medium-sized phocids that inhabited the Caribbean until recently and survive in the Mediterranean and Atlantic. Monk seals comprise the genus, *Monachus,* for which the "monachine" seal group is named, and they may resemble the seals that inhabited the subtropical Atlantic in the Miocene. They are the only phocids that inhabit tropical waters today, and seem less specialized than other living ones. They are sexually monomorphic and noncolonial, with scattered populations living near shores or islands, and feeding in relatively shallow waters.

Webb's Imperial Formation fossil apparently had migrated from the Caribbean to the Pacific, and the fact that another monk seal species lives in Hawaiian waters today seemed to support this. It's hard to see how monk seals would have reached the Pacific otherwise. When I mentioned Webb's fossil to Clayton Ray and Lawrence Barnes, however, they were surprised and skeptical. They thought it might be an odobenid fossil instead of a monk seal; Barnes and others have found odobenids in the Imperial Formation.

Northern elephant seals, *Mirounga angustirostris,* are such a presence on today's west coast that one might expect a better fossil record of them than of the rarer monk seals. Both are "monachine" phocids, which generally are larger and more southerly than "phocines" like harbor seals. It would seem reasonable if elephant seals had entered the Pacific via the Panama seaway before it closed in the late Pliocene. Miocene fossils of monachines from

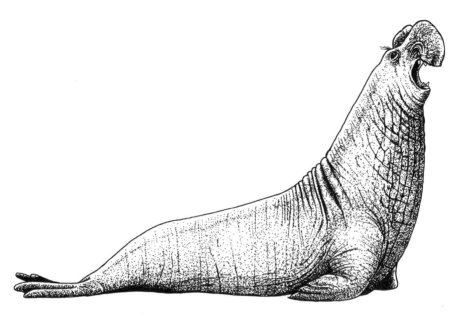

Figure 19. The elephant seal, *Mirounga,* is the largest living pinniped. The west coast species, *M. angustirostris,* attains a length of eighteen feet and can dive a mile deep, staying submerged for an hour while hunting fish and squid.

which elephant seals may have evolved are known from the Atlantic. No definite Pliocene *Mirounga* fossils have turned up on the west coast, however. A possible one from near San Diego may be an odobenid instead. And elephant seals also inhabit the Southern Hemisphere today (their genus name is an Australian aboriginal word), so they might have reached the west coast from the South Pacific instead of the Caribbean.

Another living presence may have arrived from the Caribbean, since it once inhabited the North Atlantic too and is absent from the Southern Hemisphere. Gray whales, *Eschrichtius robustus,* seem even more likely than elephant seals to have left fossils, given their size (adults typically measure forty-five feet) and habits. Unlike other mysticetes, grays hug the coast, bearing their young in Baja lagoons, staying in sight of land during migration, and feeding on bottom amphipods and worms in the shallow Bering and Chukchi seas. Large coast-hugging whales might have been expected to leave many bones in lagoon deposits like the Drake's Bay ones. But gray whales' Miocene or Pliocene past is as blank as *Mirounga*'s, although their taxonomic history is such a tangle that "blank" is perhaps not the right word.

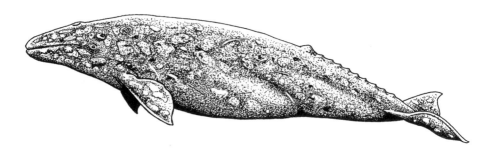

Figure 20. The Pacific gray whale, *Eschrichtius robustus,* is the west coast's commonest large cetacean, but fossils have revealed little about its evolution.

The old troublemaker Edward Cope caused much of the tangle. Gray whales' genus is now *Eschrichtius* because that was what an English taxonomist, John E. Gray, named some Atlantic gray bones unearthed in Sweden in 1864. Gray thought his new genus was one of the more streamlined and pelagic rorquals, however, because whalers had exterminated Atlantic grays in the early 1700s, and he didn't know they had existed. Soon after Gray coined *Eschrichtius,* the youthful Cope began applying the name to a miscellany of baleen whale fossils he was excitedly and untidily gathering on the east coast.

"Methods of collecting and preserving fossil cetacean remains were less than satisfactory during the past century," complained Remington Kellogg in 1968. "Inexperienced and untrained persons were picking up at waters' edge bones from the fallen debris of the marine formations. . . . The novelty and supposed rarity of such finds seemingly influenced Leidy and Cope to seek the assistance of workers in marl pits and helpers on farms in the recovery of such rarities. Broken and otherwise imperfect vertebrae and mandibles of fossil cetaceans . . . became the basis of generic and specific names."

Cope named a half-dozen east coast fossil *Eschrichtius* species and applied the name to a Pliocene jaw fragment from a San Diego well, which he considered related to rorquals of the genus *Balaenoptera* (including blue whales). But he gave the Pacific gray another name entirely. "A schooner load of bones of this species, just gathered in Scammon's Lagoon, lower California, recently arrived in San Francisco, and were sold to be ground to fertilizers," he wrote while visiting the coast in 1879. "Having examined a large number of the bones I can complete the characters of the genus, *Rachianectes,* which has been but imperfectly known."

Cope evidently saw no similarity between his *Rachianectes* and Gray's

1864 Atlantic gray whale fossil, not surprisingly, since he paid little attention to Gray's specimen. "It is now certain," Kellogg lamented, "that not one of the mandibular types of the fossil mysticetes referred to *Eschrichteus* by Cope exhibits even a remote resemblance to the mandible of *Eschrichtius robustus.*" Cope later repented of his *Eschrichtius* fling as his caution about marine evolution grew, dropping the name when he classified the Cetacea in 1890, but he kept calling Pacific gray whales *Rachianectes*. They did not get their present tongue-twisting name (after a Danish zoologist named Eschricht) until 1937, when taxonomists compared the 1864 Swedish bones with newly discovered ones from Holland and untangled the nomenclature.

Gray whales continued to trip up unwary researchers. Because the whales hug the coast and have fewer baleen plates than other mysticetes, biologists once thought them older, more "primitive," than rorquals like humpbacks or blues. Now, however, they think grays may be younger than rorquals, because their coast-hugging adaptations such as suctioning food from bottom sediments seem "derived" compared to most baleen whales,' implying that grays evolved from a more pelagic ancestor. Their baleen plates are unusually thick, also apparently a derived trait. So gray whales may have moved from the open ocean to fill a coastal niche, although nobody knows how or when.

After the late Miocene's mass extinctions, the early Pliocene must have been a regenerative time on the west coast. An incubatory warmth seems to linger at Point Resistance on sunny late winter days, when the Drake's Bay cliffs hang like laundered sheets above the water, and iris and buttercups color the hills. Along with *Thalassoleon,* the early fur seal-like otariid, the formation has produced many late Miocene and early Pliocene whale bones. They protrude from cliffs like porous tree limbs after storms.

Douglas Emlong considered Point Reyes poor fossil territory compared to Oregon, but during two days of collecting at Drake's Bay in the winter and fall of 1974, he found a cetothere jaw, a small pinniped radius, and the "nearly complete, well preserved skull of a large odontocete, possibly killer whale." Typically, he explored the formation's remotest site, the steep headlands of Drake's Estero. The skull was "in float" on the beach below a cliff, probably having recently fallen. Sandstone boulders there today have chunks of dark red bone in them, like pimentos in giant cocktail olives. I don't know how Emlong got one containing an orca-sized skull to the nearest paved road, several miles away.

The Drake's Bay Formation's pale sandstones and shales actually do come from warmer climes. They formed some eighty miles southeast of their present location in lagoons like those around the Gulf of California today, when the climate was probably as warm as northern Mexico's. Along with the rest of Point Reyes, they inched northwest in the ensuing 5 million years, dragged by the sideways interaction of the North American and Pacific plates.

Tectonic activity seems to have had a new start in the Pliocene. Subduction of Pacific crustal plates under the North American one had gone on throughout the Miocene, producing volcanoes and a gradual elevation of the continental interior, but not towering coastal mountains. In the Pliocene, tectonic interactions became more violent and complex. While some plate margins continued to subduct, others scraped against each other, tearing the coast into deep transform faults and pushing up crustal blocks at dizzying rates.

North America's tectonic turmoil didn't affect the west coast's marine life as much as it did land organisms. While faulting and uplift created rain shadow deserts and chilly high plateaus in the interior, the coast remained relatively lush and warm. Something happening to the south would affect coastal marine life more.

One of the places plate subduction dwindled by about 5 million years ago was Panama, where a seaway had divided North and South America since the Paleocene epoch. A side effect of subduction is that a deep trench forms offshore of a continental plate as it rides over an oceanic plate and forces that plate down into the earth's mantle. When subduction stopped at the Panama Strait, the trench there disappeared, and land gradually rose between the continents, forming the Panama isthmus by the mid-Pliocene. Southern California's Imperial Formation provides evidence of the isthmus's formation, because, above the marine sediments containing possible monk seals from the Caribbean, land sediments contain armadillos and ground sloths from South America.

The Panama land bridge closed the seaway that, for 50 million years, had been carrying cetaceans, sirenians, and, possibly, sea otters, seals, and gray whales to the west coast. But that probably wasn't its most important effect on west coast evolution. The seaway's closure had major, if unclear, effects on global climate. It may have engendered the Gulf Stream as it deflected a great current that had run between the continents, so that the water flowed up North America's east coast instead. Warm, moist air carried north by the Gulf Stream may have increased snowfall in higher latitudes enough to trigger North America's continental glaciation.

Whatever the exact nature of such geological and climatic interactions, their upshot would be a plunge in global temperatures unprecedented since the Permian period's great extinctions 250 million years earlier. The Pacific's vast mass of water buffered this plunge along the west coast, but, after the mid-Pliocene, the climate inexorably cooled. Temperatures still fluctuated, but the Miocene's prolonged summer would not return.

Another pulse of extinction coincided with late Pliocene cooling. West coast monk seals may have been casualties of it. The osteodontorns, the giant "toothed" marine soaring birds, disappeared, along with the flightless auk, *Praemancalla.* Odobenids made their final west coast appearance in the late Pliocene San Diego Formation near the Baja border. According to Thomas Deméré, *Valenictus chulavistensis* was "a tusked walrus closely related to modern *Odobenus*" except that, nearly toothless except for its tusks, it was even more adapted for sucking mollusks from their shells. "The toothlessness of *V. chulavistensis* is unique among known pinnipeds, but parallels the condition seen in modern suction-feeding beaked whales." The other last western odobenid, *Dusignathus seftoni,* was a double-tusked one like *Gomphotaria,* but Deméré thought from its limb bones that it might have hunted fish and squid like the earlier walrus relative, *Imagotaria.*

Cetaceans underwent some loss of diversity. An ancient mysticete group, the cetotheres, disappeared at the Pliocene's beginning, and an ancient toothed group, the platanastids, disappeared at its end. Another toothed group, which includes today's beluga whales and narwhals, had been common in west coast waters since the Miocene. According to Lawrence Barnes, the big skull Emlong found at Drake's Bay probably was that of a beluga relative, not a killer whale, being much older than any known orca fossil. The beluga group also disappeared from the west coast after the Pliocene, however, and survives only in the Arctic, perhaps because it undergoes less competition there from the more recently evolved dolphins and porpoises.

Some creatures, as usual, passed through the late Pliocene extinction unscathed. The sirenian, *Hydrodamalis,* continued to thrive, according to Daryl Domning, through its "large size, thick epidermis and blubber," and its "adaptations to feeding on relatively soft plants near the surface and in shallow and turbulent water." *Hydrodamalis cuestae,* the Pliocene species that preceded Steller's sea cow, ranged from California to Japan by 2 million years ago.

Some new organisms appeared. A new flightless auk genus, *Mancalla,* became common. According to Hildegarde Howard, its wings were more modified for swimming than *Praemancalla*'s, and it seems to have lasted until recently. Fossils from Japan and California suggest that sea lions appeared in the Pliocene, distinguished from the older fur seals by single-rooted cheek teeth and loss of dense underfur. These changes may reflect environmental shifts, although tooth simplification and fur loss have occurred since the beginning of marine mammal evolution, and they don't seem to give sea lions a competitive advantage over fur seals. Sea lions are more common on the west coast today, but northern fur seals, *Callorhinus,* dominate the pelagic niche.

Orcas may have reached the west coast in the late Pliocene, although no fossils that old have turned up here. Dolphins, of which orcas are the largest genus, evolved in the Miocene and, like porpoises, probably have occupied the west coast since then. Giant delphinoid teeth are known from Pliocene deposits in Japan and Italy. An Italian skull is from a slightly smaller animal with more teeth than living orcas, suggesting a transition from a smaller whale. The living genus, *Orcinus,* "whale from the nether-world," is the most widely distributed marine mammal.

If older toothed whales like belugas and platanastids disappeared from the west coast because orcas and other dolphins arrived, it would make sense. Dolphins have the largest brains in relation to body size of any ceta-ceans except sperm whales, which suggests, among other things, that their echolocation, sonar communication, and other typical odontocete features are exceptional. They behave accordingly, living in complex, long-lasting groups that practice sophisticated hunting strategies.

On the west coast, which provides a good habitat with its cool waters and many bays and inlets, orcas have evolved separate populations that divide up prey resources efficiently. A larger one lives mainly on fish, and its many family groups range along shore from Puget Sound north, and also around offshore islands like the Queen Charlottes. A smaller population preys mainly on marine mammals, from blue whales to sea otters, and is more nomadic. These are the orcas that gray whales try to avoid by passing close to Point Resistance on their way north with calves. Apparently the shallow water and kelp hinder the orcas' echolocation. Grays have been observed hiding in kelp beds when orcas were echolocating nearby.

The spottiness of the west coast's Pliocene fossil record might seem odd considering that it is more recent than the Miocene's. There is at least one good reason for it. The Pliocene lasted 3 million years to the Miocene's nearly 20 million, so fewer fossils had time to accumulate. But it still seems ironic that the fossil records of long-dead *Enaliarctos* and *Aetiocetus,* of plotopterids and osteodontorns, are better known than those of living ele-phant seals and gray whales, pelicans and cormorants.

We might know more today if *Aetiocetus*'s discoverer had continued collecting through the past quarter century. "As a close associate and advisor," wrote Clayton Ray of Douglas Emlong's strange career, "I inevitably speculate. 'What if . . . ?'" But Emlong ran out of time. The Smithsonian, its research funds drying up as Sputnik's memory dimmed, largely stopped supporting his work in the late 1970s. By then, he'd spent the money from his collection, partly, the story goes, on a girlfriend's drug habit. According to Ray, he also fell victim to "a fly-by-night finance company which 'contracted' his income for a small amount" to "California music publishers who promised to market his music writing efforts for large fees" and to "a book publisher who did the same for the manuscript of a book, 'Advent of Immortality.'"

A book about his fossil hunting might have sold, but Emlong avoided the subject in "Advent of Immortality," which remained unpublished. Uneasiness about his scientific status may have contributed to this. "I am both curious and jealous regarding this material," he once wrote Ray.

> I hope and pray that none of this fine material such as the Coos Bay pinniped skulls ends up on the table of measurements as comparative material not even illustrated, while far away materials get the limelight. This would make the rigors of collecting seem almost useless. Consider the amount of material I have gotten to your institute, in relation to what benefits such as [sic] California institutions may bring yours, in weighing such factors.

Emlong tired of bone hunting at times. When trying to sell his fossils in 1966, he spoke of becoming an art teacher instead. "I spent five days a week for 12 years on the beach in winter building the collection," he told a journalist, "and now I'm through with it." He never pursued art far—his few paintings are crude, albeit memorable. He was more serious about music, which he perhaps considered his true vocation. "These songs are the sum of my torment and dreams," he wrote Ray in 1977. "This music is as big a thing for me as the fossils, for it roars through my head, with the instrumental parts present."

"Advent of Immortality," in fact, is an extended rock lyric, a melange of New Age pseudoscience, zany visions, and romantic fantasies that evokes only faint echoes of marine life: "I walk for miles on the beach, beauty and cravings for love and a Spanish rapture assail me after a period of morose dysfunction. A towering ankh centers at the top of my head and extends upward like a Christmas tree." Jennie Emlong said it was about "something that to him looked very beautiful, like traveling through space," and "the fact that some of the intelligent people who lived on this earth seemed to disappear suddenly and nobody knew where they went." It goes beyond the usual "close encounter" scenarios, however. Emlong fantasized an escape from earthly limits, through space travel and biotechnology, to an undying sensual bliss:

> Beings could manipulate DNA to form animate appendages to live through other humanoid form. These could be programmed to have no ego of their own. They could also be made to undergo metamorphosis stages, like some insects do, starting with fish-like humanoid forms for underwater delights, then when wanted the humanoid form for land, and then a winged being for flight. Our oldest dreams might be implemented in this way, like a pair of lovers migrating from sea to land to air in a stay on some wild world. So much for tales of mermaids by ancients. The Babylonians, Greeks, and others spoke of a fish-like intelligent alien who gave them knowledge and help.

Emlong also tried more practical, if equally strange, ways of transcending life's limits. "I have gone through a harrowing period of emotional stress and turmoil this year," he wrote Ray in 1973, "but have developed some methods of relieving the symptoms that might even be of interest to medical science, if I can be fully successful in strengthening my own mind and body by these means." Jennie said he "spent a lot of money buying cobalt magnets and putting them together in various configurations, he called them." This involved wearing magnets on his head and other body parts. "He had a theory that if the magnet was built a certain way, that it would give a boost to your strength. . . . [He] was lifting those heavy rocks all the time that, I suppose, were too heavy for his strength."

The limits drew in on him, though. "I am recovering from what was a nervous breakdown," he wrote Ray in 1978, "but can collect fossils again now, rather than being confined to the house, as I was for some time." He still aspired to find marvelous things. "I know you will be thrilled at the fossil elephant-like animal, with purple agate teeth which could be displayed even in the gem section," he wrote after a 1979 trip to the Bakersfield area. "The unidentified mammal skull skull [sic] from Pyramid Hill will also amaze you."

In 1980, he went to California again. As Jennie told Ray:

He called and said he'd found this very fine specimen, which he hoped to sell to the Smithsonian to help finance the publishing of [his] book. Then he called again and said he didn't know if he had enough strength to dig it out. Then he called again and said that he had dug it out and got it loaded into the car, but he thought he was dying. I think then he decided he'd better try to get home with it. He got as far as Santa Cruz and the car broke down, and it was tied up in the garage for several days. That forced him to rest up there in the motel for a while so he was able to drive home. He arrived on June 7, and told me again that he was dying. But he still wouldn't go to a doctor.

Emlong's uncanny gift had failed him this time. The trip produced no thrilling specimen, only, according to Ray, a "not very promising" concretion containing "considerable broken, spongy bone." He didn't bother taking it out of his car when he got home to Oregon.

On June 9, beside a photo of hot ash blasting from Mount St. Helens's eruption, the Portland *Oregonian* ran a piece headed, strangely, "Officers See Man Push Self Off Cliff." It quoted a state trooper, who said that when he and his partner were on the Otter Crest Loop trail overlooking a high ocean cliff at 4:30 the day before, they had seen the man lying on the grass outside the guardrail, oblivious until they "stepped around him." Then: "He rolled over and looked at me and pushed himself off the cliff. We didn't talk or anything, he was just gone. I couldn't believe it was happening. There was just nothing to do. Usually when people jump, they talk first. He didn't even leave a note." The victim was "a 28 year old Lincoln City man . . . Douglas Ralph Emlong," who "apparently had been despondent recently over personal problems."

The next day, the *Oregonian* ran a longer piece, "Marine Fossils Great Passion in Life of Coastal Plunge Victim," with a photo of a smiling twenty-year-old Emlong displaying "one of his fossil finds—the head of an ancient porpoise." It was a pinniped head, but the piece did better at identifying Emlong this time, getting his age right at thirty-eight.

"He had a scientific genius that brought him acclaim," wrote the reporter, Ann Sullivan. "At least a dozen scientists from all over the world are studying Doug Emlong's fossils under the auspices of the Smithsonian. But he also led a lonely, troubled adult life—his plunge from the cliff Sunday was declared a probable suicide by Dr. Peter Cookson, Lincoln County medical examiner." The other trooper said they had run to the guardrail when they "heard moans," and found Emlong beside "an empty bottle of gin, an almost empty bottle of vodka, and evidence of 'a great many' red pills that he had evidently vomited." When the Coast Guard recovered his body, impact with the water had torn off his clothes and probably killed him outright.

"The youth prowled the Oregon coast alone, especially after winter storms," Sullivan continued. "But he was not heralded in his own state." She described the University of Oregon's 1966 attempt to confiscate his collection, and the Land Board's decision "that it would be a cruel thing to take the collection from Emlong, since it was the youth's 20 years of hard work that made it valuable." She gave the Smithsonian higher marks for heralding him, overstating the fossils' purchase price at $40,000 instead of $30,000.

Sullivan closed with an enigmatic story: "Mrs. Emlong Monday recalled her child's first showing her ammonites—extinct coiled shells of the Mesozoic era—he had discovered. And she repeated his words."

"'I think I found something different.'"

What did Emlong mean? Perhaps simply that the Oregon fossils were unlike the snakes, cactuses, and arrowheads he'd collected in California. Or perhaps he meant that they revealed an entire world different from the one wherein his mother's shaky finances had dragged them bewilderingly from place to place. As Sullivan observed, "most of Emlong's specimens, painstakingly dug out of beach sands and cliffs of the central coast, ranged from the Oligocene through the Pliocene." In that respect, they did represent a different world wherein, after Cretaceous catastrophes, a long stable period restocked an emptied planet with more life than ever.

But no world is eternal, as Emlong, despite his fantasies, had reason to know. "Whatever fleeting peace of mind that he found," Clayton Ray recalled, "seemed to come mostly through the successes of his lone search for fossils on the stormy beaches, where he also found the inspiration for his painting and composing, ineffably integrated in a way that only he could feel—but of late even these things seemed to afford him less and less respite from his personal torment." The late Pliocene's changes were foreshocks to catastrophes greater than anything in the preceding 63 million years. And although west coast life would miss some of the biggest, it was in for a bumpy ride.

FIFTEEN

Ice Age Invasions

The most spectacular seascape I've seen at Point Reyes was a morning's walk south of Point Resistance. A brushy slope hid it from the coast trail, so well that as I scrambled up I felt the top might reveal not the Pacific but the outer space through which Douglas Emlong dreamt of traveling. I was relieved to see water and sand below—somewhat relieved, given how far below they were. It was a brilliant, windy June day, and the clifftop seemed excessively exposed.

At first, the long crescent beach around the bay looked empty. Then, at the far end, I discerned what might have been a line of beige mealworms above the surf. Seen through binoculars, they were harbor seals, hundreds of them, more than I'd seen together anywhere. More kept arriving, as though magically, since I never saw them approaching through the shallows, only appearing in the surf and humping up the sand. They were perform-ing the shark-evasive maneuver of hugging the bottom until they reached the safety of the beach. There they pushed into gaps in the line and subsided.

It seemed odd that they packed themselves into one section of beach. When I crept to the cliff edge, however, I saw another line of perhaps sixty seals just below me, divided from the others by a quarter mile. The shallows below this smaller contingent contained a lot of kelp and rocks, and floating there were tiny whitish objects that twitched about fitfully, evidently baby seals playing. Perhaps the smaller contingent was a nursing colony, the larger one a mating congregation. Harbor seals have been seen mating at that beach, although, consonant with their discreet ways, only after long observation.

Point Reyes supports about a fifth of California's roughly 40,000 harbor seals and produces a larger percentage of its pups. This is not surprising. Despite its small area, it provides good habitat, and, as one of the coast's few legislated wilderness areas, has miles of roadless shoreline. But the beach congregation effect was still striking. I felt as though I was looking at a marine version of the southern Great Plains before 1870, when bison herds still grazed a sea of grass, disturbed only by the occasional grizzly or wolf.

It was not just that I was seeing hundreds of seals in a wilderness. The Pleistocene plains were different from Miocene and Pliocene ones. The bison and elk herds that amazed European explorers were absent in the earlier epochs, and, of living North American ungulates, only pronghorns and deer then inhabited the continent, although they were part of an immense diversity of ungulates. Beside multiple equid, camelid, and proboscidean species, Miocene savannas included oddities like chalicotheres, clawed horse relatives, and entelodonts, which vaguely resembled giant pigs but were unrelated to any living mammal. Pliocene prairies were decreasingly diverse, and most large native ungulates had vanished when the Pliocene ended 2 million years ago, evidently victims of climate change.

Bison and elk invaded a faunally impoverished North America from Eurasia in the Pleistocene. And that was when harbor seals invaded a west coast similarly emptied of dugongs, desmostylians, odobenids, and oddities like the oyster bear *Kolponomos.* By the time Point Reyes's little *Phoca vitulina,* "calf seal," arrived, a single sirenian genus remained, and sea lions and fur seals may have been the only nonphocid pinnipeds. Even the cosmopolitan whales were less diverse than before. The historical abundance of relatively recent arrivals like harbor and elephant seals seems primevally rich, but it was a product of primeval impoverishment.

There is a problem in comparing western land and sea faunas, however. We know fairly well what bison and elk evolved from, and how they got to North America—by crossing a Bering land bridge sometime between a million and 100,000 years ago. We still aren't sure whether harbor and elephant seals originated from otterlike animals on the Atlantic coast or from *Enaliarctos*-like ones on the Pacific. If the former is so, it remains unclear how phocids got from the Atlantic to the North Pacific. If the latter, it remains unclear not only how phocids got to the North Pacific, but how their Oligocene ancestors got to the Atlantic.

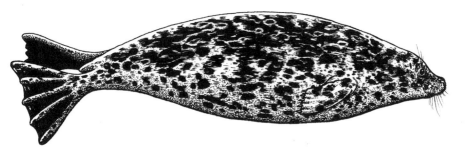

Figure 21. The harbor seal, *Phoca vitulina,* "calf seal," is the smallest and most commonly seen pinniped on the west coast today. Nobody knows exactly how this relatively recent arrival got here.

If marine mammal evolution generally has been hard to understand, phocid evolution has been particularly so. Phocid fossils have been even more elusive than other pinnipeds', and early paleontologists had the same problems with them that Remington Kellogg complained about in their whale studies. They seized on ill-assorted fragments of ancient bone, named them with abandon, and drew shaky parallels between fossil seals and living genera.

The harbor seal's genus, *Phoca,* is an example of such confusions. When Jeffries Wyman, the anatomist who had debunked the "sea serpent," *Hydrarchus,* described some skull, leg, and vertebral fragments from a Virginia Miocene formation in 1850, he likened them to the bones of the living harp seal, *Phoca groenlandica.* Joseph Leidy, the Philadelphia anatomist, used his description to designate a new species, *Phoca wymani,* also attributing a canine tooth from the same formation to the species in 1857.

A decade later, Edward Cope attributed Leidy's canine tooth to a cetacean named *Squalodon,* not a seal. He was right, and Leidy accepted the correction, although *he* attributed the tooth to a cetacean named *Delphinodon.* It was a small error, and Cope didn't question the rest of Leidy's *Phoca* species. Still, it occurred soon before Leidy failed to advise him about the head-to-tail plesiosaur gaffe that so amused Marsh. With the "fossil feud" in mind, later writers assumed that Cope had invalidated Leidy's entire *Phoca wymani,* not just a tooth, although none studied Wyman's fossils, which gathered dust in Harvard's Museum of Comparative Zoology.

Paleontologists continued to name fossil Miocene seals *Phoca,* a practice that, as Clayton Ray wrote, had the effect "of influencing taxonomic and evolutionary interpretations through the power of nomenclature." By naming ancient seals *Phoca,* in other words, they convinced themselves that living *Phoca* is an ancient genus. The sense of inevitable progress that made Marsh and his successors identify his *Desmostylus* as a flippered dugong influenced this. Marsh's toothed-bird assistant, Samuel Williston, while disliking the Yale professor's lordly airs, saw evolution similarly. When he became a professor, Williston proclaimed that "the largest sea animals have been the final evolution of their respective races."

Termed "Williston's Law," this sweeping dictum prevailed for a century. In 1881, William Flower, who noted affinities between pigs and whales, wrote that the elephant seal, *Mirounga* had "in the fullest degree all the characters by which the Seals are distinguished from the terrestrial Carnivora." In 1966, Edward Mitchell wrote that sexually dimorphic, pelagic *Mirounga* was "the most specialized and advanced phocid." By that standard, small, sexually monomorphic, coast-hugging *Phoca* was primitive, so it seemed logical to associate the earliest phocid seal fossils with it.

Again, however, the "higgledy-piggledy" of Darwinian natural selection doesn't mandate change in any particular direction. If the marine desmostylians reproduced successfully for millions of years with four legs instead of flippers and flukes, why couldn't a seal genus do so while getting smaller instead of larger? One paleontologist, André Wyss, suggested in a 1994 paper that "large size is most likely the ancestral condition" of phocids, and that the smaller size of genera like *Phoca* is "best regarded as secondary."

Wyss acknowledged that "[l]arge body size among marine mammals is generally considered an adaptive response to the physiological demands imposed by a heat-dissipating aquatic environment," and that "the trend among otariids and odobenids, for example, seems to be toward increasing body size." But he thought from skeletal traits in fossil and living seals that the smaller seals might be "neotenic," reaching reproductive age in a more juvenile form than their ancestors. This is a not uncommon phenomenon (humans have neotenic traits compared to extinct hominids), although not a well understood one. Wyss's paper did not explain seal neoteny, and it cautioned that other traits of small seals like *Phoca* are the opposite of neotonic. Still, neoteny might explain one possible circumstance of an early Pacific to Atlantic migration, if phocid seals are descended from an *Enaliarctos* relative.

Organisms in stressful environments sometimes reproduce at increasingly early ages, one possible cause for neotenic evolution. Known fossils suggest that the earliest phocids evolved in a subtropical North Atlantic and adjacent seas. Early seal bones in the southeastern United States are associated with crocodiles, sirenians, and other warm water animals. If, as Wyss and Annalisa Berta have maintained, the early west coast pinniped, *Desmatophoca,* had more traits in common with phocids than with otariids, it might have shared an ancestor with other early pinnipeds that ranged far enough south in the Pacific to pass eastward through the Panama seaway. There, finding a warm, nutrient-poor South Atlantic stressful compared to a colder, nutrient-rich Pacific, that ancestor's descendants might have gotten smaller and acquired other phocid traits, like loss of ability to rotate their hind limbs forward. Wyss has observed that phocids generally display neotenic traits compared to other pinnipeds.

Unfortunately for this theory, the known fossil record does not contain any *Desmatophoca*-like pinnipeds from the Atlantic. The earliest apparent phocid seal fossils, the South Carolina Oligocene leg bones used to support the diphyletic, otterlike ancestor version of pinniped origins, "are definitely more closely related to the Phocidae than to other carnivores," according to Irina Koretsky and Albert Sanders. They found the bones "closely comparable to the most specialized phocid, the modern genus *Cystophora,*" the large, Arctic hooded seal, although they weren't suggesting that the Oligocene seal was closely related to *Cystophora,* only that the leg bones are similarly specialized.

However phocids evolved, the fossil record now shows that both monachine and phocine seals lived in the Atlantic by the early mid-Miocene, but that they were quite different genera from living ones. When Clayton Ray examined Jeffries Wyman's dusty Virginia Miocene seal bones at Harvard in the 1970s, he found that they were indeed from a phocid, but not from *Phoca,* as Leidy had thought. He tentatively identified them as from an extinct genus, *Monotherium,* "*Monachus*-like beast," except for a tooth and a leg bone, which he attributed to another extinct genus, *Leptophoca,* "thin little *Phoca.*" So Leidy's original *Phoca wymani* had been a composite not only of a phocid seal and a small whale, as he and Cope had thought, but of a monachine seal as well.

The larger monachine seals like *Monotherium,* well adapted to subtropical conditions on both sides of the North Atlantic, seem to have predominated in the mid-Miocene, so they may indeed have evolved first. The smaller phocines like *Leptophoca* may have originated in more marginal habitats such as isolated remnants of the south Asian Tethys Sea, perhaps becoming more neotenic because conditions were more stressful than the open ocean. The smallest living seals, the genus *Pusa,* resembling dwarf harbor seals, inhabit a still-shrinking Tethyan remnant, the Caspian Sea.

In the mid- to late Miocene, both kinds of phocid seals became numerous enough that they spread into the South Atlantic and South Pacific, as African and Peruvian fossils show. It is strange that they apparently didn't reach the North Pacific then, since they may have entered the South Pacific by passing through the Central American seaway. But there are no known west coast Miocene phocid fossils. Maybe the beaches were too crowded with otariids and odobenids. Or maybe phocid fossils simply haven't turned up.

There is a west coast fossil of a Pliocene phocine seal, a flipper bone found in southern Alaska. It is attributed to a small genus like *Pusa* or *Phoca,* and nobody is sure how it got there, especially if such genera evolved in Eurasian inland seas. One theory is that, since they live in fresh as well as salt water today, the seals may have swum into the Siberian rivers that feed lakes like Baikal, and followed them to the Arctic Ocean. The several species of *Phoca* seals live all around the Northern Hemisphere today, however, so a river migration explanation may be too involved. The little Pliocene phocine simply may have moved from Atlantic to Pacific via a "northwest passage."

In any case, the obscure little phocine's North Pacific arrival heralded complicated developments in the subsequent Pleistocene epoch. Beginning about 1.8 million years ago, unrelievedly cold, wet weather caused continental glaciers to spread over higher latititudes at least four times, while ice sheets formed on even tropical mountain ranges. The weight of ice pushed continental crust hundreds of feet into the Earth's mantle and erased life from much of North America and Eurasia. Plant fossils show that tundra prevailed as far as Spain and Ohio. Forests and lakes covered today's deserts. Pack ice covered the Arctic Ocean for the first time.

Bizarre things happened. Volcanoes erupted under ice sheets, flooding vast areas with cascading meltwater. When continental glaciers retreated, subglacial meltwater lakes suddenly emptied, flooding even vaster areas. During interglacials, climates were sometimes warmer and drier than today's, and even during glaciations, faunas were exotically diverse by modern standards. Pleistocene sites on the west coast yield jaguar and tapir fossils as well as mammoths, lions, and cheetahs. Given the geological and climatic upheavals, it's not surprising that faunal migration increased.

Ice sheets reached the west coast shoreline only from the high coastal mountains of British Columbia and southern Alaska, but glaciation affected it drastically by lowering sea level hundreds of feet. The shoreline moved dozens or scores of miles farther west than at present, sometimes to the edge of the continental shelf. This would have made life harder for many coastal creatures by erasing shallow water and bay habitat. It also would have closed the Bering Strait and stopped marine migration to and from the Arctic Ocean. It is hard to imagine how some present species like gray whales lived under those conditions. Baja lagoons and Bering Sea shallows where they now breed and feed would have been miles inland.

Gray whales did inhabit the Pleistocene west coast, however. One very like the living species left its bones on the Palos Verdes Peninsula near Los Angeles during an interglacial period between 50,000 and 120,000 years ago, when sea level was a little higher than now. So grays evidently were migrating from California lagoons to Bering Sea feeding grounds during that interglacial. They might have migrated even farther north when the Bering Strait was open, if some similarly aged ear bones discovered on Alaska's North Slope are from a gray whale. But since these are the only gray whale fossils known, the species' activities during the last glaciation, from 50,000 to 13,000 years ago, or during other glaciations and interglacials, remain as mysterious as its origins.

The past of the west coast's other commonly land-hugging cetacean, the harbor porpoise, is even murkier than *Eschrichtius*'s. Although the porpoise family began leaving fossils in the Miocene, the harbor porpoise has left none that we know of. Like the gray whale, it may have existed before the Panama land bridge formed, because it occurs in the North Atlantic as well as the North Pacific. And like the gray whale, it may be a fairly derived species. It has an unusual life cycle for a cetacean. Calves are weaned very early, within five to seven months of birth, and individuals reach sexual maturity in little more than a year instead of in two like most cetaceans. These traits, and the species' small size and shortened, bulbous cranium, seem indications of the neotenic process that may have influenced phocine seal evolution. Perhaps the stresses of ice age life contributed to this, but so far there is no evidence.

The fossil record says no more about the harbor seal's appearance on the west coast than the gray whale's or harbor porpoise's. "Phocines responded to deteriorating climates in the North Atlantic after the early Pliocene perhaps by retreating southward but also by adaptation," Clayton Ray wrote. "Sometime between then and the late Pleistocene when the living species appear, the modernization of the phocine skeleton occurred, in my opinion coincident with increasingly severe extremes of seasonal climate and expansion of the icy Arctic."

It may be that the harbor seal, which is not specialized for Arctic ice, reached the west coast early in the Pleistocene, when the Bering Strait was open and before glaciation was well underway. *Phoca vitulina*'s coast-hugging ways seem almost a throwback to the aboriginal Oligocene pinniped, *Enaliarctos*. But, as the gray whale shows, apparently primitive traits can be recently derived ones. In one way, harbor seals may be the most derived pinnipeds of all. Unlike others, they can breed and raise their young entirely in water. Harbor seal pups swim at birth and can stay underwater for two minutes when two days old. It seems paradoxical, since they are less pelagic than other pinnipeds, but harbor seals may be evolving toward a more fully aquatic way of life than their relatives'.

Rather than being more primitive than their ice-adapted relatives, indeed, harbor seals may have evolved from ice dwellers. They don't occur significantly earlier in the fossil record than Arctic seals like *Phoca hispida,* the ringed seal. At one site on Prince of Wales Island in southeastern Alaska, harbor and ringed seal fossils occur together with bones of Steller's sea lion, *Eumetopias jubata*. The fossils, carried by scavenging carnivores to a limestone lair called "On Your Knees Cave," date from between about 13,000 and 24,000 years ago, which corresponds roughly to the last glacial maximum. The ringed seal bones, which predominate, suggest that conditions were arctic, with pack ice at least during the winter, since that is ringed seal habitat today. Arctic fox fossils in the cave support this. Yet harbor seal and sea lion fossils occur "throughout the ice age range of On Your Knees Cave deposits including the last glacial maximum."

As the Prince of Wales Island bones show, the invading phocid seals were not the only pinnipeds to prevail through the Pleistocene's upheavals. Sea lions and fur seals continued to thrive in the Pacific and South Atlantic, although they never seem to have reached the North Atlantic. Other fossils show that the living sea otter, *Enhydra,* appeared and spread all the way around the North Pacific Basin, and that Steller's huge sea cow, *Hydrodamalis gigas,* did the same. Walruses, the last odobenids, somehow occupied the extreme North Pacific and, like phocine seals, adapted to living with ice.

The Pleistocene also recapitulated the old story of an ursine creature enticed into the water by abundant food. In this case, the creature is recent enough that fossils and molecular studies furnish a good idea of its ancestry. In mid-epoch, Eurasian brown bears that had been living on arctic shores began to specialize in catching seals on the ice pack. In the best higgledy-piggledy Darwinian style, they adapted to this new environment not only by evolving white fur but by traveling hundreds of miles over the

Arctic Ocean with the same four legs as their land ancestors. The polar bear was first named *Thalarctos maritimus,* "marine sea bear," because it spends so much more time in the water than other bears. But it swims like other bears and is now less redundantly called *Ursus maritimus,* "marine bear."

It may seem strange that a bear adapted to sea life for less than a million years should be such an effective predator on animals adapted to it for at least 30 million years. Having lost the omnivorousness of their land relatives, polar bears are the most formidable carnivores on the planet today, commonly killing walruses and beluga whales as well as seals. Males do not hibernate, spending the winter on the ice pack, where they stalk prey resting on the ice or rising at breathing holes. Summer is their lean time, since melting ice cuts them off from prey, and they may go four months without a meal. On the other hand, polar bears and their prey have been adapting to the icy Pleistocene for a similar amount of time. If the glacial epoch was a time of disaster and impoverishment, it was also a time of evolutionary creativity.

Mammals were not the only marine tetrapods to undergo radical Pleistocene change. Fossils show that murres, auks, and guillemots, confined to the Pacific since their appearance in the late Miocene, invaded the North Atlantic, perhaps drawn by cooling water temperatures and increasing fish populations. Gulls and terns, having left very few fossils in earlier epochs, appeared in all their present diversity and abundance.

While some avian giants like the "toothed" osteodontorns disappeared, others thrived. In the 1930s, bones of a giant, flightless duck relative began turning up in late Pleistocene marine terrace deposits on southern California's Channel Islands. Hildegarde Howard speculated in 1947 that the species, *Chendytes lawi,* had become flightless because it was confined to the large islands that appeared when sea level fell drastically. Subsequent discoveries on other Channel islands "substantiated this hypothesis by providing not only added evidence but a presumably ancestral species, *C. milleri* Howard (1955), from lower Pleistocene deposits that showed an earlier stage of the degenerative trend in the pectoral complex." Combined with the marine location of its fossils, unusually large neck vertebrae suggested that *Chendytes* might have foraged in the water off reefs and shores, reaching down to pluck food from the bottom.

Another giant seabird survived until historic times. While marooned on Bering Island, Georg Steller encountered a flightless cormorant that he described as a "special kind of large sea raven with a callow white ring around the eyes and red skin about the beak, which is never seen in Kamchatka."

The birds were common, unwary, and meaty, weighing up to fourteen pounds, so the expedition ate a lot of them, baked in clay to allay their toughness and fishiness. Unlike his sea cow, Steller's "sea raven" may have lasted a century or two longer in western Aleutian waters. Natives reported the existence in the mid-1800s, before firearms came into use, of a seabird "fully twice as large as the red-faced cormorant and of different plumage." The birds had disappeared by the 1880s, and all that remains are four skins and a resounding name, *Phallacrocorax perspicillatus,* coined by Pallas, the Russian naturalist who edited Steller's notebook. It's now known as Pallas's cormorant.

Late Pleistocene disappearances heralded the advent of yet another invader. Fossils suggest that *Hydrodamalis gigas* lived from Baja California to Japan until the last glacial maximum, so its confinement to the remotest North Pacific when Steller found it is circumstantial evidence of some recently lethal factor. Pinnipeds probably didn't prey on *Hydrodamalis,* and even predators large enough to tackle it—orcas, sharks, bears—must have had trouble. Steller likened sea-cow hide to tree bark. A new factor was required to decimate such a tough old survivor. As with other Pleistocene invasions, evidence of this factor long remained unknown because the ice age coastline lies under a resurgent sea. But some has begun to emerge.

SIXTEEN

Hands into Paddles

A bridge across a slough at Drake's Estero on Point Reyes is a good place to watch leopard sharks swim in and out with the tide. As I approached it one day, a sea lion that had been lying on the abutment dived into the water with a startling splash. The apparition evoked Norse fairy tales about trolls under bridges, but as the creature surfaced and eyed me apprehensively, I knew who the troll was. Humans have little to fear from sea lions, but we must seem devilish to them—unpredictable, prepotent wraiths on the borders of their world. The one under the bridge was wounded, although its wound looked more like a slash or bite than a bullet hole.

The encounter again raises the dark side of a coastal mythos. The sea creatures that came ashore in my dream were joyful and triumphant, but, as I've said, when one does so in waking life it is often a victim of accident, disease, or predation, and probably doomed. I wonder if these creatures have a sense of land as the "undiscovered country from whose bourne no traveler returns" that we have of the sea. Whales, with their big brains, probable long memories, and virtually complete separation from land life, might especially sense this. There is something fatally ritualistic about their mass strandings, a mysterious phenomenon that probably has existed a long time. Twenty-five-million-year-old fossil whales in South Carolina are lined up and scavenged by fossil crabs as though they'd died together on a beach.

For at least the past 10,000 years, humans have played a deadly role in west coast sea creatures' return to land—slaughtering disabled castaways, pursuing the strong and healthy even in their breeding grounds. Of course, humans have admired and respected sea creatures, too. Prehistoric cultures generally were forbearing—or inefficient—in exploiting them, compared with profit-mad historic ones. Still, the energy that humans have brought to killing animals often very difficult and dangerous to reach is fearsome. Necrophilic mystiques attach to whaling, and attitudes toward lesser creatures have been as bloody-minded. An Oregon coast native told me that when the state tried to reintroduce sea otters there in the 1970s, local fishermen killed them, even though they knew they eat invertebrates, not fish. Whether or not that was the main cause, sea otter reintroduction failed in southern Oregon.

Uncertainty about humanity's biological relationship to the sea complicates our ambiguity toward its creatures. We are primates, an order that, according to mainstream zoology, has no truly aquatic, much less marine, form. Our closest living relatives, the great apes, fear to cross water, a rare trait among mammals, most species of which swim readily. Some monkey species wade, swim, or forage on beaches, but that is mainly because the water is near the forest they inhabit.

Still, living humans are an exception to most anthropoids. Many can swim, some better, at least, than sirenians. Many spend much of their lives on the sea. And this has been going on for many thousands of years, coupled with equally ancient traditions of marine humanoids—nereids, tritons, mermaids, were-seals—traditions that did not necessarily seem incredible to early biologists like Georg Steller. He perhaps was speaking figuratively when he called the strange creature he described off Alaska a "sea ape." Relatively little was known about primates in the 1740s, however. Naturalists would not identify gorillas for another century. If bearlike or piglike creatures could have flippers and flukes, why not an apelike creature?

More recently, human differences from apes have convinced some evolutionists that our past might have been partly marine. According to a British writer, Elaine Morgan, scientists have had the idea at least since 1912, although she credits a later article by an Oxford marine biologist with bringing it to public notice:

> In 1960 Professor Sir Alister Hardy suggested that there must have been
> an aquatic phase in human evolution. Such a phase would account for
> a number of human characteristics unique among primates but common
> in aquatic mammals: loss of body hair, subcutaneous fat, face-to-face

copulation, the position of fetal hair, and so on. He also suggested that an aquatic phase could explain the preadaptations necessary for the emergence of bipedalism and speech. There was massive sea-flooding in northeast Africa during the so-called "fossil-gap"—roughly eight to four million years ago—after which the bipedal humans first appeared. The aquatic hypothesis suggested that flooding induced isolated populations of apes to adopt an aquatic life style. When the waters subsided, those apes would have been forced back to land.

Hardy conceived his hypothesis in 1930 after noting a similarity between seal and human fat during a polar expedition. He concealed it before 1960 for fear of "professional suicide" and didn't publish on it afterward. But Morgan continued to argue that prehominids underwent a marine aquatic phase. "[A]s the descendants of water-dwelling apes returned to terrestrial life they had small heads, rounded limbs, and smooth black skin," she wrote, an arresting if imaginative image, since she could cite no fossil evidence for it. She maintained, however, that fossils of an Italian Miocene ape "preserved in large numbers by sinking into mud" demonstrated a marine primate's possibility. She speculated that the Italian ape, *Oreopithecus,* became aquatic as sea level rose around islands it inhabited, and that this "paralleled conditions obtaining in northeast Africa during the fossil gap, at the time when, according to the molecular biologists, the evolutionary divergence between humans and apes first began."

Morgan also cited, as evidence of aquatic affinity, early hominid bones found with turtle eggs and crab claws, and bones of more recent *Homo erectus* with deepwater shellfish. Otherwise, she discounted paleontology, quoting a major evolutionary theorist, Ernst Mayr, to the effect that "[i]deas based on a study of comparative anatomy have in no case been refuted by subsequent discoveries in the fossil record." This, she maintained, supported her contention that "undiscovered fossils . . . cannot be allowed to apply to closure of aquatic debate."

Morgan's dismissal of paleontology was peremptory, however. Her own fossil examples of aquatic affinity were dubious. A well-known paleontologist expressed long-standing opinion in describing her "marine" ape, *Oreopithecus,* as "arboreal, judging from its environment, forested marshland," while crab claws and turtle eggs could as well have come from river deposits as marine ones. And Morgan chose a fatally anachronistic example of paleontological shortcomings when she wrote in her 1984 *Oreopithecus* article that there was "no fossil evidence to confirm the theory" that whales "had a land dwelling ancestor." Evidence already existed, albeit sparsely, and the detailed whale evolution scenario that emerged contrasted sharply with her vision of apes becoming marine as sea level rose, then returning to land as it fell. None of the Eocene's known four-legged marine whales returned to land as the Tethys Sea shrank. Surviving whales became more marine.

Most scientists reject Hardy's sea ape hypothesis, and it does contravene known evolutionary trends. There may be exceptions such as snakes to the rule that going to sea is a one-way trip for tetrapods, but there is no unequivocal proof of them. It is hard to see how a relatively recent marine arrival like the sea otter could return even to freshwater, because *Enhydra* is no longer adapted to the streams its ancestors inhabited, and the river otter, *Lutra,* occupies those now. The only description I've seen of sea otters in rivers and lakes is Steller's account of Bering Island, where river otters were absent. Such barriers to return must always have existed. In the Miocene, west coast rivers were full of a hippolike rhinoceros named *Teleoceras* whose prevalence might have excluded desmostylians from surviving a deterioration of marine habitat by moving into estuaries and streams.

Fossil apes with clear marine adaptations such as flippers are unknown, as are fossil apes, which like desmostylians, have been found in marine deposits consistently enough to imply a seagoing life despite a lack of flippers. The same goes for hominids before *Homo.* Early genera like *Australopithecus* seem to have had diets like chimpanzees, relying on plant foods, opportunistically scavenging or hunting animal ones. Chimpanzees live beside the Rift Valley's Lake Tanganyika today without catching fish.

The public has been less skeptical of Hardy's hypothesis than scientists, partly because of Morgan's eloquent advocacy, and partly for the same reason that it likes legends of seals come ashore and transformed into humans. Indeed, Morgan's advocacy seems as much an evolutionist retelling of the legends as scientific theorizing. On a superficial level, a human ancestor that dived among coral reefs, eating fish and oysters, compares favorably to one that trudged across savannas, gnawing nuts, bugs, or the occasional unwary baboon. On a deeper level, a human ancestor rising from the surf evokes the unity of life.

Even if it had existed, Hardy's sea ape would not explain living humans' marine affinity. Neither he nor Morgan suggested that aquatic apes began to make rafts, paddles, or other cultural adaptations to aquatic life. The scientific rationale for these, of course, is that they are like fire, buildings, and clothing, part of *Homo*'s million-year-old cultural breakout from African woodland and savanna. The sea ape hypothesis might have implications for cultural adaptations to the sea. Vestigial marine biological abilities—swimming, salt tolerance—might have facilitated later making of rafts and fishing with tools, an kind of "inheritance of acquired traits" à la Lamarck and Cope. Whether and how such traits could actually influence cultural evolution is another question, however. An animal that swims naturally might not build boats.

To say that human marine adaptation is cultural rather than biological doesn't explain it. In fact, it makes it more mysterious, since there is little evolutionary precedent for human cultural adaptation aside from tool-using animals like sea otters, and little evidence of how it occurred. The deepwater shellfish Morgan mentioned in association with *Homo erectus* fossils might suggest that our genus somehow learned to swim and dive fairly early in its evolution. But raft and boat origins remain a virtual blank, except that they probably occurred by at least 50,000 years ago, because *Homo sapiens* reached otherwise inaccessible Australia then. One reason for this blank is that rafts and boats presumably originated during the last glaciation, when sea level was much lower, so that most sites involving them or other cultural marine adaptations are now underwater.

The question of human arrival on North America's west coast is part of the mystery of early human sea-going, although the idea that prehistoric people might have gotten here in boats seemed unlikely at first. When a Jesuit historian, José de Acosta, suggested in 1588 that "savages and hunters" had been the first arrivals, he probably assumed that they had walked across from Eurasia, although not enough was known about North America for him to specify. Steller, noting similarities between Siberian and Alaskan natives during the Bering expedition, seems to have been the first to explicitly suggest an Asian origin for Americans.

Subsequent thinking fastened on the Bering Strait as a former land bridge, and Louis Aggasiz's mid-nineteenth-century discovery of continental glaciation and falling sea level supported this. From 1895 to 1903, one of early anthropology's most ambitious research projects attacked the problem as Franz Boas, Henry Fairfield Osborn's colleague at the American Museum of Natural History, sent collectors to the Pacific Northwest and Siberia to find evidence of cultural links. They found so much evidence that Boas never got around to publishing his conclusions, but a Bering land arrival was largely taken for granted afterward.

To be sure, the ice age predominance of mountain glaciers throughout southern Alaska raised an obstacle to walking between the continents. But it seemed possible that lower elevations of semiarid central Alaska had been ice-free during the last glaciation, which ended some 13,000 years ago. "Savages and hunters" might have followed the Yukon and MacKenzie rivers to the Canadian Rockies' east slope, then diffused through the hemisphere.

In the 1920s and 1930s, the discovery of big-game hunting artifacts at Folsom and Clovis in New Mexico and at Fell's Cave in Argentina buttressed land-crossing assumptions. Inhabited between 10,000 and 11,000 years ago, the sites were until recently the oldest reliably dated evidence of humans in the Americas. In the 1960s, a University of Arizona paleontologist, Paul S. Martin, proposed that humans had traversed an ice-free corridor from Siberia around 12,000 years ago, spreading rapidly to Patagonia by exploiting large land mammals unaccustomed to human predation. Opposing theories, based on possibly older artifacts, had humans occupying the hemisphere more slowly, using a range of subsistence strategies besides big-game hunting. Despite the controversy, a land crossing remained the dominant paradigm. Although some archaeologists thought migration might have been coastal as well, most assumed that southern Alaska's glaciation had extended to the sea. Humans thus would have occupied the west coast as a side branch of the main migration corridor down the Great Plains.

The Bering land-crossing paradigm weakened in the 1980s, however. No 12,000-year-old human fossils or artifacts turned up in central Alaska or northwest Canada to support the ice-free corridor route theory. Geologists who studied the distribution of glacial moraines and erratic boulders along the northern Rockies' eastern foothills began to doubt that such a corridor could have existed before 11,000 years ago. And human artifacts possibly over 12,000 years old turned up at Monte Verde, a site near the southern Chilean coast. The artifacts—wooden structures and tools, stone choppers and projectile points—do not necessarily reflect a marine way of life, since they are associated with land animals and plants. But they don't rule one out, since they are less specialized for land hunting than Folsom or Fell's Cave artifacts.

Geologists began to doubt that glaciation would have blocked a coastal route. Since the ice age coast lay many miles offshore of the present one, researchers extracted underwater sediment cores to explore the sunken terrain. They found 16,000-year-old pollen and seeds off the Queen Charlotte Islands, 30 miles west of British Columbia, and a 10,000-year-old land plant root at a depth of 300 feet 10 miles north of Vancouver Island. This suggested that much of the southern Alaskan and British Columbian coasts had been ice-free well before human remains appeared in the Americas. No sites as old as Folsom or Monte Verde have turned up on the Northwest coast, but some are almost as old. In On Your Knees Cave on southeast Alaska's Prince of Wales Island, which yielded seal and sea lion bones, researchers also found worked obsidian and fragments of the skull and pelvis of a man who had died between 9,000 and 10,000 years ago. Carbon isotope analysis of the bones suggested "a diet extremely high in fish, sea mammals and other marine organisms, as opposed to terrestrial animals and plants." Stone artifacts recently found off the Queen Charlotte Islands may be more than 10,000 years old.

A site off the California coast may be as old as Folsom and Fell's Cave. In the 1940s, Phil Orr, an archaeologist at the Santa Barbara Museum of Natural History, found chipped stone on Santa Rosa Island, one of the Channel Islands. A woman's arm and leg bones turned up in Arlington Canyon on Santa Rosa in 1959, and these may be 12,000 years old. Possible evidence of 10,000-year-old human occupation exists at a cave site on San Miguel Island, 4 miles from Santa Rosa Island.

If humans occupied such sites between 9,000 and 12,000 years ago, the implications are significant, because Santa Rosa and Prince of Wales islands stood miles from the mainland even then. Unless they were outstanding swimmers, the people who left bones and artifacts there would have needed boats to reach them. And if people had boats then, coastal immigration from Asia seems likelier than once thought. Even if glaciers did block the shore in places, immigrants could paddle around them. According to one author, Tom Koppel, coastal immigration would have had just about every advantage over land—fewer insects and predators, more food, warmer temperatures. Koppel and others proposed a migration scenario that seems a coastal version of Paul Martin's land one, with a wave of invaders exploiting unwary marine faunas to spread rapidly from the Alaskan coast to the Chilean.

"In this way, people could have populated the narrow coastal strip right down to southern Chile in something like one or two thousand years, with hardly any serious push into the interior of either continent. Rapid coastal migration could account for the 12,500 to 13,000 year old sites like Monte Verde." One striking bit of circumstantial evidence for this scenario is Steller's sea cow, which probably remained common all along the west coast until the last glaciation ended. Given its easy accessibility and superior edibility, it could have been a major inducement to southeastward coastal migration, with the immigration vanguard staying in one place long enough to exhaust the sea cow population, then moving on to the next.

Yet sea cows also pose a problem for a rapid coastal migration scenario. Given the wholesale extinction of American land mammals within the last 10,000 years—mammoths, mastodons, ground sloths, horses, camels—sea cow survival to historic times even in one place seems anomalous. And it is not the only survival that seems so. Rookery-breeding marine mammals and birds are very vulnerable to humans with boats, as historic extinctions have shown, but few marine tetrapods are known to have become extinct on the west coast since the last glaciation. Aside from Steller's sea cow and Pallas's cormorant, two other flightless seafowl—the auk, *Mancalla,* and the gooselike *Chendytes*—are the only evident ones. Prehistoric people may have wiped out pinniped rookeries on the mainland and more accessible islands. Still, pinnipeds and seabirds bred abundantly on many islands when whites arrived, and sea otters often basked on beaches.

The abundance of marine animals, along with their pelagic habits when not breeding, may have protected them. Prehistoric hunting bands would have had trouble making a dent in a rookery of tens or hundreds of thousands. Bison likewise might have survived through sheer numbers. Paul Martin once joked that land immigrants might have lived on bison while they wiped out mammoths and horses. Coastal immigrants might have lived on pinnipeds, salmon, and shellfish while they wiped out sea cows and flightless seabirds.

Still, compared to the grand scale of land extinctions, the overall survival of marine megafauna doesn't buttress a blitzkrieg coastal immigration scenario. And a blitzkrieg was not required. Small bands of hunter-gatherers could have diffused down the coast to southern Chile without exterminating their prey, especially if it took them 2,000 years rather than the 600 that Martin postulated for a land blitzkrieg. And there could have been even more time if the human fossil record in the Americas proves older than currently believed.

On the other hand, there might have been logistical problems with a brief, hemisphere-long coastal migration, as with a land one. Coast-hugging people from northern Asia would have found familiar habitats as far as California, but then things would have changed. Desert, then tropical forest, would have replaced tundra and conifers on shore; reef fish and sea turtles would have replaced salmon, seals, and sea cows in the water. (Dugongs were extinct from the tropical Americas by the Pleistocene: manatees inhabited only the Atlantic side.) A coast like the Northwest would not have turned up again until southern Chile. Whether cultures accustomed to a North Pacific coast way of life would have adapted to such changes in even a thousand years is a good question.

It is not a question that can be answered with available evidence. Imperfections of the human fossil record are even more enigmatic than with other creatures, since the thoughts and feelings they evoke are more complex. The possible early remains on the Channel Islands captured Douglas Emlong's imagination, for example. He went to the Santa Barbara area in 1974 and promptly found human bones and teeth at the base of a beach cliff. "This site is very puzzling and interesting to me," he wrote Clayton Ray, "and I hope something can be done about it as soon as possible." Ray thought the bones looked recent, and when he sent the material to an anthropologist, he got no response. Emlong's Santa Barbara fossils probably will remain no more than puzzling.

Despite coastal-migration advocates' argument that the proof must be underwater, the fact remains that no vestige of a boat, harpoon, or other definitely aquatic contraption has turned up from 12,000 to 10,000 years ago in western North America. Such artifacts do not appear until much later, and then they appear abruptly, as though the people who made them came from the earth or the water. And according to many of their descendants, earth or water was where they came from.

SEVENTEEN

Pileated Woodpecker's Boat

Whether the first immigrants arrived in boats or not, several kinds existed on North America's west coast when European explorers arrived. Sebastián Cermeño, a Portuguese captain who visited Point Reyes in 1595, reported that a man met his ship in "a small boat of grass which looks like the bull-rushes in the Lake of Mexico." The man, evidently a Coast Miwok, sat in the middle with a two-bladed paddle. These tule rafts, along with others made of logs tied together, were the only boats the Coast Miwoks had, and they apparently didn't go far out in them. They are not known to have eaten sea mammals despite their abundance on the Farallones.

Rafts seem primitive, and it is tempting to see them as the first watercraft that might have plied the west coast 12,000 years ago. People used them along hundreds of miles of California, and although the users differed in many respects, most had languages and other traits suggesting that they'd been in America a long time. Some linguists have classed Coast Miwok in a group called Penutian, which may include Central America's Mayan languages. One linguist, Joseph Greenberg, classified most languages of west coast raft-users in an even larger group called Amerind, which would include all Central and South American languages as well as most North American ones.

Greenberg recognized two other major language groups in North America, Na-Dene and Eskimo-Aleut, and, interestingly, speakers of those languages were major users of two other kinds of west coast boats. In southeast Alaska, the Tlingits, speaking a Na-Dene Athabaskan language, used dugouts carved from logs, and other dugout-using coastal groups south to California spoke Athabaskan languages. Dugouts allowed fairly long sea trips, and some groups, including the Tlingits, hunted gray whales. In western Alaska, the Eskimo-Aleut-speaking Yupiks, Aleuts, and Eyaks used boats of skin sewn around bone or wood frames. These allowed them to go farther and stay out longer than dugouts, and they hunted bowhead whales as well as grays.

According to Greenberg's theory, Na-Dene and Eskimo-Aleut languages entered North America later than Amerind, and geography would support this, since they are confined to the north and west. It is tempting then, since dugouts seem more advanced than rafts, and skin boats more advanced than dugouts, to interpret the three kinds of watercraft as a north-to-south progression, with Amerind speakers arriving in rafts, Na-Dene speakers later arriving in dugouts, and Eskimo-Aleuts arriving still later in skin boats.

It is not quite that simple, however. To the south of many raft users, along the Santa Barbara Channel and on the Channel Islands, an Amerind-speaking people called the Chumash used boats made of sewn-together planks caulked with tar, which seem more advanced than at least rafts or dugouts. Some were large, and their makers used them to hunt, if not whales, then smaller marine mammals, as well as swordfish and tuna.

The Chumash plank boat has long been a puzzle. When Cermeño saw some near the Channel Islands in 1595, he described them as "like the baru-tillos of the Philippines," but most anthropologists later thought the Chumash had developed it here, from a raft or dugout. More recently, the idea that it came from the eastern Pacific as Cermeño implied has gained momentum. Polynesians used sewn-plank boats, which occurred in Peru along with Polynesian cultural aspects such as sweet potatoes. California has no physical suggestion of Polynesian presence besides the boats, but a Chumash word for them, "tomolo'o," resembles a Hawaiian word, "kumulaa'au," for the redwood drift logs from which they made planks. And Chumash plank boats seem to have appeared around 1,300 years ago, when Polynesians were making Pan-Pacific explorations.

Other complexities beset a theoretical progression from Amerind rafts to Na-Dene dugouts to Eskimo-Aleut skin boats. West coast boats are not necessarily the key to human arrival in the hemisphere. Many groups throughout the Americas used boats, including rafts, dugouts, and skin boats as well as types unknown on the west coast, like birch-bark canoes. Some archaeologists, observing that Clovis artifacts are more like those of the European Solutrean culture than any found in Siberia, have suggested that America's first boats might have arrived on the Atlantic coast instead of the Pacific.

Possible human sites in Pennsylvania and South Carolina may be older than known western ones.

Even on the west coast, language groups did not neatly define boat use. Many dugout-using groups, like the Haidas, who inhabited the Queen Charlotte Islands, south of the Tlingits, spoke Amerind languages—or possible Amerind languages, since some of those, Haida included, have been hard to classify. Some Athabaskan-speaking groups, like the Cahtos and Wailakis, of northwest California used rafts.

At least one west coast group reportedly did not use boats, the Coast Yukis of northern Mendocino County, California. They were unusual in other ways. They spoke a language different from other Native American groups and are said to have been smaller and longer-headed, traits resembling some of the continent's oldest human remains, like "Kennewick Man" in Washington State. If the Yukis were descended from some of the earliest immigrants, this might suggest that North America's first people did not arrive in boats.

As with gray whales and harbor seals, evidence of human presence on the west coast is too sparse and complex for much scientific certainty. Of course, there is also the evidence of what the aboriginal coastal people themselves have had to say, although that tends to be sparse and complicated too. José de Acosta, the sixteenth-century Jesuit who first theorized on prehistoric immigration from Eurasia, called it "a matter of no importance to know what the Indians themselves report of their beginnings, being more like dreams than histories." Still, dreams can contain information.

The reportedly boatless Coast Yukis exemplify such evidence's sparseness and complexity, and also its tantalizing existence. According to available records, about 750 of them occupied 50 miles of coast in 1850, living in shell-mound villages at stream mouths, subsisting on acorns, salmon, shellfish, and grass seeds supplemented by game, which probably included basking pinnipeds and otters, along with the occasional beached cetacean. By 1864, their population had dropped to 50 because of disease, starvation, and homicide attendant on white immigration. In 1910, the census was 15; in 1926, it was 4. In 1972, two anthropologists, Alfred Kroeber and Robert Heizer, dubbed them "ethnologically extinct," meaning no survivor was "culturally Indian" or "spoke the language."

By chance, however, we know more about how the Coast Yukis regarded the sea than other groups south of Oregon. "Coast Yuki mythology has many marine characters, as might be expected from the littoral habitat of these people," wrote another anthropologist, Edward Gifford, in 1937. "No doubt various coast peoples of California had mythologies with similar characters, but unfortunately these have remained largely unrecorded." Much of the knowledge came from an old man named Tony Bell who told Gifford some stories in 1926. They were called "night stories" because they were supposed to be told only on winter evenings, although Gifford heard them in English during the day.

"The events told in the night stories occurred before the appearance of the present race," Gifford wrote. "The characters of the stories constituted the prehuman race, which the Coast Yuki believe preceded them." All of Bell's night stories involved the sea. A brief one said that humans emerged from a rock in the ocean near the present town of Rockport. Tunnels in the rock ran in the four directions, and when a god heard singing from the rock, he told them to come out. Another said that Yukis originated from an old couple named Sand Flea (the crustaceans that swarm in surf sand), who stayed behind after the prehuman race waded away across the ocean. A longer story, interestingly, involved building a boat for a long-distance voyage. Called "Pileated Woodpecker Crosses the Ocean," it was traditionally told if the winter rains hadn't begun by the solstice.

The long story's woodpecker hero participates in the world's origin by teaching the animals how to feed themselves, but gets bored when they disperse and leave him alone with his two wives, two sons, and father-in-law: "After a time, the chief thought of making a boat (alaltat). He went and cut a stick separately and made a boat in the assembly house. The women never saw him. One day he finished it. He intended to cross the ocean. He fixed red feathers for a crest and white feathers for a neck ornament. He was all dressed up (mixnik). He did all this in secret so his wives never saw it. He intended to take Nuthatch (or Creeper) with him across the ocean, so that Nuthatch would talk to him and tell him where to land."

Woodpecker takes his wives mussel-gathering, then sends them home: "Just before they got to the top of the rise they looked back and saw him far at sea with his red crest waving. He had launched his boat very quickly. The women watched the receding boat until they could see it no longer, weeping bitterly all the time."

The voyage is Odyssean:

After a time, the chief came to a bank of kelp which he could not pass through at first. He persevered and finally succeeded in making the passage. Then he encountered a succession of obstacles in the ocean: a bank of surf fish, a bank of bull heads, a bank of blue cod, a bank of red rock cod, a bank of seals, a bank of sea lions, a bank of blind (or flying) whales. It was getting late when he passed through the whales. He next encountered his greatest difficulty—a bank of sharks. He had a terrible time passing through them. Next he passed a bank of salmon. Still he was not close to land. Moreover, Nuthatch, riding on his head, had not spoken yet.

Nuthatch finally tells him to land at "another father-in-law's house," where Woodpecker finds a second family and tries to continue his deceitful behavior. His second father-in-law tells his new wives to show him some salmon in the creek the next morning, but having already seen them, he sneaks out early to catch some. When he takes them to the house, however, the father-in-law dismisses them as inedible, then secretly eats them himself.

The next day, one of the wives takes Woodpecker to the stream to show him how to catch the salmon properly, a startling experience:

> She told the chief to watch. "When I throw a stone," she said, "those fish will come. Don't gig those on top but watch for a white-gilled fish at the bottom. You gig it." The fish was really a whale. She threw a stone and the fish poured in. Again she admonished him to be sure to gig the white-gilled one on the bottom. He gigged it, but he could not hold it. His wife had to catch him, his harpoon, and the "salmon." She struck the "salmon" with a stone fish club. The "salmon" was really a whale. The chief did not know how to carry it. His wife said "you must take it home." He tried but could not lift it. Then she carried the whale home on her finger.

The father-in-law cooks the whale and serves strips from the fins to the chief and his wives. But Woodpecker has had enough of this disconcerting new family and decides to make a "return voyage" after challenging his father-in-law to a rainmaking contest. The old man fails to produce a storm, but Woodpecker succeeds and, gratifyingly, almost drowns his powerful relative. He then embarks with Nuthatch and travels home.

There he finds his original family, but he has been gone a deceptively long time. They collapse in surprise at his return, and, although he revives them, things fall apart: "When he saw his two wives, he decided to live with them in the assembly house. After some time, the women decided to take their sons and go away. Their father did not know of this plan, but Pileated Woodpecker did. They departed and only Woodpecker knew where they were going. Then Woodpecker went away by himself and is going yet. The women are going yet."

Judging from Bell's stories, the Coast Yuki considered human origins, perhaps all life's origins, marine, and ocean voyaging a part of their past. They called themselves Ukoht-ontilka, "ocean people." It is unclear, however, whether this reflects traditions thousands of years old or simply the fact that, as Gifford wrote, such ideas "might be expected from the littoral habitat of these people." It is even unclear whether the stories substantially represent Coast Yuki beliefs. Pileated Woodpecker's boat, which sounds like a dugout canoe, made from "a stick separately," might be borrowed from other tribes' stories. Kroeber wrote that the Yuki knew of canoes "as used by tribes to the north." Tony Bell might have made up at least parts of the stories from his own experience and imagination.

"The writings of many ethnographers, both popular and scholarly, are well supplied with sentences in the form 'The Haida believe that . . .' or 'The Navajo believe that . . . ,'" a Canadian writer, Robert Bringhurst, has observed. "Perhaps not all such sentences are altogether false, but it is certain that no such sentence is ever entirely true." Taking Bell's stories as typical of Coast Yuki, much less Native American, attitudes toward the coast might be like taking Thoreau's transcendentalist travelogue, *Cape Cod,* as representative of New England, much less Euro-American, attitudes.

Yet some stories Bringhurst has translated from Haida suggest that myths like Tony Bell's did have deep roots on the west coast. Few coastal groups were farther apart, geographically and culturally, than the Coast Yukis and the Haidas, who numbered 12,000 in 1800 and were like other Pacific Northwest tribes in being rich enough from marine wealth to create elaborate settlements without relying on farming. Visiting Alaska in 1879, John Muir, an inventor as well as a naturalist, and not prejudiced in Indians' favor, rated Tlingit building skills above white immigrants'. Whites generally considered them a cut above other Indians and still did when I hitchhiked through British Columbia in 1972.

The Haidas also resembled their northwestern neighbors in being aggressively acquisitive. If their power had extended far enough south, they probably would have treated Coast Yukis as minor trading partners or slaves. Although status and remoteness exempted them somewhat from murder and starvation, however, disease hit the Haidas hard. About a thousand survived in 1900, when an ethnographer, John Swanton, visited the Queen Charlottes to record stories for Boas's Bering Strait project.

Given their marine expertise, the Haidas might have been expected to tell more sea stories than the Yukis, and Swanton's informants did. Still, similarities between Coast Yuki stories and Haida stories are striking. Both link human origins with coastal rock formations. Both tell of origin from tiny shore creatures, although cockle snails take the place of Tony Bell's beach fleas in a Haida version. Both describe heroes' voyages to dangerous marine realms where everyday scale and proportion no longer apply, and where superhuman inhabitants are related to the voyagers. Both tell of "fish" that, when caught, startlingly prove to be whales.

"Pileated Woodpecker Crosses the Ocean" shows the greatest similarities with a longer story called "Raven Traveling" that an old Haida man named John Sky told Swanton in 1900. Both stories feature big, noisy land birds that take long sea trips, and both bird-men have dual natures. They are heroes who participate in origins, tricksters who try to get more than their share of sex, food, and wealth while evading work and responsibility. Both stories begin with origins, although "Raven Traveling" goes into more detail:

> Hereabouts was all saltwater, they say.
> He was flying all around, the Raven was,
> Looking for land that he could stand on.
> After a time, at the toe of the Islands, there was one rock awash.
> He flew there to sit.
> Like sea-cucumbers, gods lay across it,
> Putting their mouths against it side by side.

Raven flies until he finds villages, where he enters an infant's body and sneaks out at night to eat villagers' eyes. Caught because of his gloating by "someone who was half rock, living in the back corner," he is thrown into the ocean, where he wallows self-pityingly until a voice tells him his father wants to see him. When he looks through the eye of his marten-skin blanket, he sees a pied-billed grebe who repeats the invitation and dives. A "two-headed kelp" turns into a house pole, and he climbs down it:

> It was the same to him in the sea as it was to him above.
> Then he came down in front of the house,
> And someone invited him in.
>
> "Come inside my grandson.
> The birds have been singing
> That you would be borrowing something of mine."
> Then he went in.
>
> At the back sat an elder, white as a gull,
> Who asked him to bring a box that hung in the corner.

When the old man had it in his hands,
He took out five boxes, one inside the other.
In the innermost box were cylindrical things:
One that glittered
And one that was black.

As he handed him these, he said to him,
"You are me.
You are that, too."

On top of the screens forming a point at the rear of the house,
Sleek blue beings were preening themselves.
Those are the things of which he was speaking.

The white elder shows Raven how to make the land by upending one cylinder in the water, biting half off the other one, and spitting it at the first. The hero's first attempt fails when he tries to do it backward, but he eventually gets it right:

Trees came into being then, they say.
When he laid this place into the water,
It stretched itself out.
The gods swam to it, taking their places.
The same things happened with the mainland,
When he set it in the water round end up.

Both the Yuki and Haida heroes embark on rakish wanderings but, guided by quieter, wiser birds, unexpectedly find divine relatives. (According to Bringhurst, the "sleek blue beings preening themselves" are iridescently plumaged young ravens after their first molt.) Both receive instructions for acts appropriate to their divine natures and partly bungle them. The rest of "Raven Traveling" recounts duplicity similar to Woodpecker's, as Raven tries to get Junco and Steller's Jay to guide him on a voyage involving a plot to abandon his wife and seduce his sister. Both heroes end up the same—"going yet," away by themselves.

The heroes' dual natures and disconcerting experiences reflect ambiguity toward the sea as an enricher of life but an imposer of limits. Both heroes set out, cleverly and egotistically, to get things from it. The Neptunian elder they encounter is also the god of wealth. The instructions he gives Woodpecker, via his daughter, involve the biggest windfall the Coast Yukis could hope for, a beached whale. Those he gives Raven involve the Haida version of a full bank deposit box—obsidian and crystal cylinders in nested cedar chests. But both heroes get more than they bargained for as the wealth redistributes itself unexpectedly.

The stories reveal no final authority or outcome—the sea god himself can be half-drowned by a clever trickster—only a continual recycling—a "going yet." They do imply a moral: that culture's tricks don't prevail more than temporarily, sooner or later leaving ruined voids that nature must refill, with the sea as the great refiller—a continuum of creative flux more than a source of wealth.

The similarity between the Indian stories' mythic past and the deep time of evolutionary stories is tantalizing. The teeming banks of sea creatures in Bell's story and the gods, "like sea-cucumbers," in Sky's seem to echo prehistoric life. "There are many sea-serpent stories in west-coast native folklore," Paul LeBlond and Edward Bousfield wrote in their *Cadborosaurus* book. "The representations of sea monsters in native petroglyphs are striking in their consistency." Tribes around Vancouver Island told of eight-foot-long "serpents" with long hair on their heads and backs.

Fossils seem to have influenced some sea monster stories, as with the one Cope noted when collecting on the Plains. The Pawnees and Kiowas, who lived on the Niobrara Chalk, had water serpent stories related to mosasaur fossils. Fossil legends are less common in west coast lore, not surprisingly, since coastal fossils are rarer, but people must have encountered bones and shells. The Achumwai, who inhabited the eastern Klamath Mountains, where Merriam found ichthyosaurs and thalattosaurs, told of "giant water dragons," one of which they considered ancestral to the alligator lizard, a living species common at Point Reyes. After a 1995 wildfire that burned everything in sight of Point Resistance, I found a young lizard perched phoenixlike on a charred tree.

Coastal people certainly would have recognized similarities between fossils and living creatures, which may have influenced their stories that humans emerged from coastal rocks. Stories that life began as tiny beach creatures like amphipods and snails suggest a kind of evolutionary thinking. Of course, the stories' full meanings for prehistoric people are unknown to us. Still, as they invoked the ambiguities of exploiting the sea, the stories would have helped to sustain a very old world—until another world landed on top of it.

EIGHTEEN

The End of the Earth

The historical discovery of North America's west coast was definitely by boat, although exactly whose boat is unclear. The coast remained just about equally remote from all the world's expanding civilizations until surprisingly recent times, more remote than Africa, South America, or even Australia. That made its discovery peculiarly vague. Polynesians, Chinese, or others may have been the first arrivals, but the Spanish were the first well-documented ones. Hernán Cortés sent expeditions to Baja California, leading one himself in 1535, but Indians repelled them, having heard Mexico's fate. Others crept up Baja in ensuing years, and in 1542 two ships captained by Juan Rodríguez Cabrillo, who had served under Cortés, reached Alta California. Although willing to trade, the natives were unfriendly. At what is now San Diego, they told Cabrillo that mounted men with crossbows were killing people in the interior.

Continuing north, Cabrillo saw many large houses and canoes on the Channel Islands but found the climate discouraging. This was during a pulse of cold weather, begun in the late Middle Ages, called the little ice age. Snowy peaks loomed around Monterey Bay in November, and winter storms plagued the expedition. It got as far as Cape Mendocino, then a storm drove it back to the Channel Islands, where Cabrillo broke his leg during an Indian attack and died of gangrene. His pilot, Bartolomé Ferrer, sailed north again but got no farther than the present Oregon border, where a worse storm forced him back to Mexico.

The next recorded expedition was not as well documented as the Spanish ones; indeed, it was even more mythic than Vitus Bering's Alaska odyssey. Steller's missing journal of that is a simple problem compared to the documentary tangle surrounding Sir Francis Drake's famous west coast sojourn in the summer of 1579, during his combined global circumnavigation and privateering campaign against the Spanish. The English Crown sequestered the expedition's records when it returned, and only a bewildering mélange of rumors and pamphlets dribbled out in subsequent years.

It wasn't until almost fifty years later that Drake's nephew published a more or less official account. Modestly entitled *The World Encompassed,* it is a curious book. Although authored by the nephew, also Sir Francis Drake, it was "collated" by Francis Fletcher, the expedition's chaplain, who had probably kept a journal of it, although no copy survives. Like most Elizabethans, Fletcher is shadowy, and some historians consider *The World Encompassed*'s account of the west coast a fabrication. It probably is, to some extent. It is also full of lively details, some of which suggest the west coast.

The book's narrator, whoever he is, paints a bleak but not necessarily fantastic picture of a North Pacific wherein the ship's rigging grew so stiff with cold and damp after they left the tropics that the sailors could barely handle it. Seeking a Northwest Passage, they sailed to the 48th parallel, as far north as Vancouver Island, but the sea's roughness discouraged them from venturing farther. Even the land looked ominous:

> [T]he neerer still we came unto it, the more extremities of cold did sease upon us. The 5 day of June, we were forced by contrary winds to runne in with the shoare, which we then first descried, and to cast anchor in a bad bay, the best roade we could for the present meete with, where we were not without some danger by reason of the extreme gusts and flurries that beate upon us, which if they ceased and were still at anytime, immediately upon their intermission there followed most, vile, thicke and stinking fogges, against which the sea prevailed nothing, til the gusts of wind again removed them, which brought with them such extremity of violence when they came, that there was no dealing or resisting against them.

The narrator attributes this "insufferable sharpnesse" to the "large spreading of the Asian and American continent," from whose "high and snow-covered mountaines, the North and North-west winds (the constant vistants to those coasts) send abroad their frozen nymphes, to the infecting of the whole aire. . . . Hence comes the general squalidness and barrennesse of the countrie; hence comes it, that in the middest of their summer, the snow hardly departs even from their doores, and is never taken away from their hills at all." Describing the harbor 10 degrees farther south, where they stopped to careen their ship in preparation for sailing to the East Indies, he bursts out: "[H]ow unhandsome and deformed appeared the face of the earth it selfe! Shewing trees without leaves, and the ground without greennes in those months of June and July."

Many historians think that harbor was Drake's Bay at Point Reyes; others that it was some bay in northern California. Others doubt that the English could have sailed so far north in their leaky, heavily laden little ship, much less to Vancouver Island, and suspect they invented the story to obscure the Spanish claim on California. "It seemeth that the Spaniards hitherto had never beene in this part of the country," declares the narrator, "neither did ever discover the land by many degrees to the southwards of this place." This is fabrication, since Drake used captured Spanish charts to navigate north from Mexico. Still, the book's description of the place and its inhabitants shows haunting similarities to Point Reyes.

Even in these warmer times, the melodramatic descriptions of the summer coast's "squalidness and barrennesse" have a certain authenticity. After I sat for twenty minutes beside the channel wherein Drake is thought to have careened his ship one sunny June morning, my hands were semiparalyzed from the surprisingly chill wind sweeping down the estuary. It might well have seemed to carry "frozen nymphes, to the infecting of the whole aire."

The inhabitants, according to anthropologist Robert Heizer, used words that may have been Coast Miwok. Naked except for skins and rush aprons, they were deeply disturbed at the expedition's appearance and made long, impassioned speeches, presenting the English with a staff crowned with a sphere of black feathers and hung with shell money. They returned much of the cloth and trinket presents the English offered them, and during ceremonial visits to Drake's fort on shore, women stood on the cliffs wailing and scratching their faces until they bled. Some were so upset that they didn't eat, and Drake "was faine to perform the office of a father to them, relieving them with such victualls as we had provided for ourselves, as Muscles, Seales, and suchlike, wherein they took exceeding much content."

If the place was Point Reyes, and the people were Coast Miwoks who had never seen a European, Drake's ship must have seemed straight out of a story like "Pileated Woodpecker Crosses the Ocean." They might well have thought the pallid beings in their whalelike boat and bizarre costumes were ancestors who, for some mysterious, not necessarily propitious, reason, had returned from the sea. The Miwoks had dances during which they impersonated the dead and women bled from the mouth. And mussels and seal meat do seem more likely gifts from seagoing ancestor gods than cloth and beads.

Some native behavior was less reverential. On especially cold days, they "used to come shuddering to us in their warm furres" and try to get under wool cloaks. Some tried to adopt the youngest, handsomest Englishmen as their particular friends. On the other hand, English attempts to show that they were not "gods" by conspicuous eating and drinking failed, and, according to the book, the natives held them in awe until they left, weeping and imploring them to stay. Indeed, they approved when Drake assumed that their speeches were appeals to become English subjects, and posted a bronze plaque taking possession of their land. This last tale seems more anti-Spanish propaganda, and the plaque has disappeared, but Heizer dug quantities of sixteenth-century porcelain and iron spikes from Miwok sites around Drake's Bay.

The narrator's account of the land around the harbor is also suggestive. Venturing inland from the "unhandsome and deformed" shore, they found a "goodly country, and fruitfull soyles, stored with many blessings fit for the use of man," which could be Point Reyes inland from the fog belt. These included "the company of very large and fat Deere which we saw by the thousands, as we supposed, in a hearde," which sound like the tule elk that live around Drake's Bay today (reintroduced in 1998 by the Park Service). Because the headman's cloak was made from its skins, the narrator describes in particular detail "a strange kinde of Connies," of which the "whole Countrey" was "a warren." It had a head like an English cony's, feet like a mole's, a tail like a rat's, and food-carrying pouches on either side of its chin. Some historians have cited this as fabrication, since "cony" usually applies to rabbits, and American rabbits don't burrow. But it is a fair description of the pocket gopher, and Point Reyes is a warren of its burrows. The aplodontia, a burrowing rodent that also occurs there, is another possibility.

On the other hand, the Drake book describes little marine life, which seems odd. Aside from the "Muscles" and "Seales" given to distraught Miwoks, it refers to local sea creatures three times. It mentions the cold's effect on nesting birds, an observation perhaps drawn from raids on rookeries like Point Resistance: "The poore birds and foules not daring (as we had great experience to observe it) so much as once to arise from their nests after the first egg layed till it, with all the rest, be hatched and brought to some strength of nature, able to help itselfe." It describes the natives' fishing skill: "One thing we observed in them with admiration, that if at any time they chanced to see a fish so neere the shoare that they might reach the place without swimming, they would never, or very seldome, miss to take it." And it mentions islands, presumably the Farallones, where in one afternoon the departing ship gathered enough food to get them to the East Indies, but says only that the islands had "plentifull and great store of seales and birds."

Such perfunctory descriptions might seem further evidence of fabrication. Yet sparse reference to marine life is typical of early accounts. That of the 1542 Cabrillo expedition mentions dogs, parrots, and large herds of pronghorns, but no sea animals except fish. In 1595, Sebastián Cermeño enthused about the land at Point Reyes, calling it "fertile as far as three leagues inland" and predicting: "The soil will return any kind of food that may be sown, as there are trees which bear hazelnuts, acorns, and other fruits of the country, madrones and fragrant herbs like those from Castile." But, although he had plenty of time to observe marine life after a storm wrecked his galleon and forced him to return to Mexico in an improvised launch, he recalled only that the Miwoks ate "a quantity of crabs" along with wild birds and deer, and that fish were the Channel Islands' staple food.

In 1603, when authorities sent another merchant captain, Sebastián Vizcaino, to chart the Alta California coast so wrecks like Cermeño's could be avoided, he also showed more interest in land animals than marine ones. He listed "many game animals, such as stags that look like young bulls, deer, bison, very large bears, rabbits, hares, and many others," but, although noting that seals were abundant, did not distinguish different kinds. He was more interested that the Monterey Bay Indians prepared the sealskins they wore "better than how it is done in Castille."

A possible reason for this is that mainland and near-shore island rookeries were rare. Grizzlies and coyotes would have preyed on them, and human populations were relatively large. There is evidence that the Chumash increasingly exploited pinnipeds after they began making plank boats, but then later turned more to fishing, suggesting they had depleted accessible Channel Island rookeries. An archaeological dig at what is now a large pinniped rookery on San Miguel Island revealed the site of a large Chumash settlement. On the other hand, even if rookeries had been remote, pinnipeds, sea otters, and cetaceans must have been common sights. Perhaps explorers simply were less concerned with observing marine life than with evading their own marine deaths from diseases, storms, or other threats.

Like those that it preceded it, the 1603 Vizcaino expedition found plenty of perils to evade. It encountered the usual frightening weather, and most of the crew had scurvy by the time they turned back at Cape Mendocino. Such experiences discouraged European exploration for the next century. Spanish plans to develop Monterey as a way station for the Manila trade evaporated. A 1683 attempt to colonize Baja failed after two years. California remained so little explored that geographers, extrapolating from the Baja Peninsula's dual coasts, considered it an island until the nineteenth century. Only the inquisitive Jesuits, who began to establish Baja missions in 1697, got a foothold before the Franciscans finally colonized Alta California from inland in 1769.

Miguel del Barco, a Baja Jesuit mission priest from 1737 to 1768, described west coast marine creatures at some length, but he too gave more space to land ones. Of seabirds except pelicans, he said only that there was "an excessive number of species, whose names I cannot get right." Following prevalent opinion, he classed cetaceans as fish. Some of his descriptions were secondhand, as with that of "the rarest fish, which occasionally has been seen on this coast . . . the *mulier* fish or nereide." Another priest, Victoriano Arnes, who'd found one on a northern Baja beach, said it had an ordinary fish's lower body and a woman's upper one, with a white face, neck, shoulders, and breasts, although the carcass was too dried to discern hair, nipples, or arms. "The size, as far as I could tell, was more than two *cuartas*," Barco wrote, "but that one may not have been full grown." His manuscript includes a drawing of a big-eyed, big-breasted "nereide" with a scaly back, a spiny dorsal fin, and a bow-shaped tail fin.

Barco gave more immediate accounts of other creatures. He probably made one of the first references to Pacific gray whales when he noted an abundance of "*ballenatos*" for which a canal, gulf, and cove were named. Seamen called them "*ballenatos*," he explained, because, "although very big, they aren't as big as the ones they say there are in other oceans," a fair comparison of grays with sperm or blue whales. He perceptively distinguished sea otters, which Father Sigismundo Taraval had first reported from Cedros Island halfway up the Peninsula in the early 1730s, from other aquatic mammals: "[A]lthough some have called them beavers, they aren't really, at least not like those of Canada which have the rare gift of building their own houses, since they live only in the sea. Others call them [river] otters [*nutrias*], but they aren't that either, because of their fatness, their short legs, and their nearly black color. . . . But that is the prevalent name." He noted that otters ate only shellfish, lived in the Pacific north of about 29 degrees latititude, and were ungainly on land, coming out of the water only to sleep or sunbathe.

"Also on both coasts of California," he continued, "are marine wolves which like otters are amphibious, living partly in the sea and partly on the beach. They are bigger than otters, with shorter, coarser fur. In some places, especially in some islands in the gulf, there are admirable multitudes of these wolves."

Barco's attention often wavered back to hearsay, however. He told how twenty otters killed on Cedros Island by Taraval's men had made good hats when sent to Mexico. He described an imaginative use for sea lions: "When a woman fears a miscarriage, a great remedy for preventing it is to put in her girdle, next to her body, a band or tape of about three or four finger's breadth, cut from the skin of the marine wolf."

Georg Wilhelm Steller seems to have been the first European naturalist to take an unwavering interest in west coast sea creatures *per se*. And he did so almost by default. Born in Bavaria in 1709, Steller studied enough zoology at German universities to work as a surgeon, but, like most naturalists then, was more enamored of botany. As Leonhard Stejneger, his biographer, observed, "[F]aunistic studies comparable to the contemporaneous botanical disciplines were almost unknown." Plants were easy to collect, new species were numerous, and botany fit into the systematizing theories then popular. Large vertebrates, especially marine ones, offered few such advantages. Pioneer taxonomist Carl Linnaeus's *Systema Naturae* entry on pinnipeds as "a dirty, curious, and quarrelsome tribe, easily tamed and polygamous; fat and skin useful," stands out from his tidy plant descriptions.

If Steller had been lucky and cautious, he might have emulated Linnaeus, who, only two years older, ended up with a rich wife and a secure teaching job. But, for all his sharp intelligence, Georg Wilhelm seems to have been a protoromantic, longing for the remote and sublime. Young Linnaeus made a pioneering exploration of Lapland but returned to Sweden after six months, ready to convert adventurous celebrity into professional prestige. Young Steller launched himself headlong into the unknown and never returned.

Without job prospects in crowded Germany, Steller joined the Russian army as a medical assistant and traveled on a troop transport to St. Petersburg, where he ingratiated himself with a Professor Johann Amman of the Russian Academy of Sciences. Amman certified that Steller was "not only well versed in the fundamentals of botany and the other branches of natural history, but had also shown a rare diligence in the exploration of plants." But Steller also made enemies, and an unwise marriage to the fetching but penniless widow of an older naturalist named Daniel Gottlieb Messerschmidt.

Before dying, Messerschmidt had stimulated his young colleague's wanderlust with tales of Siberian exploration. In 1737, Steller's contacts got him a kind of job as an adjunct naturalist to Vitus Bering's expedition in north-

Figure 22. The California sea lion, *Zalophus californianus,* is the "performing seal" of circuses. Sea lions evolved on the west coast in the Pliocene epoch, from 5 to 2 million years ago.

eastern Siberia, but first he had to get there. His wife had promised to go along, but she got only as far as Moscow before retreating to St. Petersburg, where she sold his furniture to augment the stipend he'd granted her. After crossing vast Siberia, Steller caught up with the expedition in 1739 but found it in disarray because of squabbling personnel and its commander's listlessness. This was Bering's second Far Eastern expedition, and he had realized during his first that he disliked exploring. He had been struggling fitfully for years to obey imperial orders and get on with a planned voyage to North America. When Steller presented himself at headquarters in Okhotsk in August 1740, the commander told him he was not authorized to include him.

Steller was resilient, however. Linnaeus and Johann Gmelin, another Siberian explorer, described him as "a born naturalist and botanist" whose outstanding characteristics were his "inquisitiveness and irrepressible habit of asking questions." After his disappointing interview with Bering, he rushed off full of curiosity to see a whale and porpoise that had washed ashore about 27 miles from Okhotsk. The whale was too buried in sand to examine, and the porpoise was an already described species, but Steller would have better luck with fish. He was the first to describe the anadromous life cycle of Pacific salmon, and he named five of the extant species, although his assistant took credit after his death.

Steller decided to give up on wild regions temporarily and botanize in Japan, observing that the North Pacific's obscurity was not surprising, since civilized people didn't notice their own countries. Then his luck changed. Bering, whose health was failing, and who liked the young naturalist, offered him a job as his personal physician on the North America voyage. Steller accepted eagerly, joining the expedition on the Kamchatka Peninsula, but then made himself a nuisance by complaining to the authorities about the army's treatment of natives, the inadequate medical provisions for the voyage, and the other expedition naturalists' incompetence. Naturalists, officers, and crew were shunning him by the time the expedition sailed in June 1741, and a month later they ignored his excited pronouncement, based on sightings of "the Kamchatka sea beaver, or more correctly sea otter . . . which lives entirely on shellfish and crustaceans and is therefore compelled to keep close to the shore," that Bering's ship was nearing Alaska.

"Because I was the first to announce it," Steller fumed, "and because forsooth it was not so distinct that a picture could be made of it, the announcement, as usual, was regarded as one of my peculiarities." He was even more frustrated when Bering forbade him to explore an offshore island they'd encountered (now Kayak Island) for fear of hostile natives, and he became so insubordinate that a less tolerant commander might have put him in irons. Bering finally let him go ashore with the watering party (heralded by an ironic trumpet blast), where he rushed about for ten hours, snatching up birds, plants, and native artifacts. One bird, the Steller's jay, proved they had reached North America, since Steller recalled a painting of its congeneric, the Carolina bluejay, by an American colonial artist. Such discoveries so enthralled him that the men had to force him back onto the ship by threatening to maroon him.

Bering then decided to turn around, and Steller might have come back from the voyage with little more than a day's land collection and shipboard observations of sea creatures. Bering was not interested in flora and fauna, only in charting the coast as ordered, and he was not very interested in that. When scurvy broke out later in August (Steller having missed a chance to collect antiscorbutic herbs during his breathless hours ashore), the officers agreed to forget charting and hurry back to Kamchatka. But fog and headwinds delayed them along the Aleutians for the next month. Then came the late September storm that wrecked the *Saint Peter* and marooned them for a year on Bering Island.

Steller showed markedly less enthusiasm for that island than for offshore Alaska, probably wishing himself anywhere except that dank expanse of maritime tundra. "The island is a succession of barren cliffs and mountains joined one to the other," he wrote. "Separated from one another by numerous valleys running south and north, they stand aloft from the sea like a single rock." But his stay virtually marked the beginning of marine mammal zoology, indeed of general field zoology—the systematic description of life histories as opposed to miscellaneous bestiaries like Miguel del Barco's.

One of Steller's responses to the earlier "sea ape" sighting off Alaska shows how far he had to go from Barco's traditional approach. Lacking a specimen of the strange creature, or any kind of field guide, he tried to identify it from the available literature: "As for its body shape . . . it corresponds in all respects to the picture that Gesner received from one of his correspondents, and in his *Historia animalium* calls *Simia marina danica*. At least our sea animal can by all rights be given this name because of both its resemblance to Gesner's *Simia* and its strange habits, quick movements, and playfulness."

Conrad Gesner, a Swiss physician and naturalist, had died in 1565, but almost two centuries later his multivolume *Historia* was still the major compendium of known and rumored animals. It has a chapter on *Simia marina*, characterizing them vaguely as *non piscis quid haec, sed bestia cartilaginea*, "not fishes as such, but cartilaginous beasts." The strange animal Steller described, however, bore no resemblance to the *Simia marina* of which Gesner's correspondent, a Dane named Kellerman, had sent him a picture. That picture shows a creature with dorsal, pectoral, and pelvic fins and a long ratlike tail fin; Gesner's text describes it as *viridis toto corpore*, "green all over." It looks like a chimaera or ratfish, a chondrichthyan related to sharks and rays. It is unlikely that Steller had the massive work with him, and he must have forgotten what *Simia marina danica* looked like, as he forgot the name of Mark Catesby, the colonial artist who had painted the Carolina bluejay.

The same page, however, shows another "sea ape" with a doglike head, apelike torso, and fishlike lower half, which Gesner labels *Serpentum Indicum*. It resembles Steller's "new animal" enough that it more likely was the source of his sea ape identification, although it doesn't shed much light on his strange sighting either. With forelimbs and without external ears, unlike Steller's creature, Gesner's *Serpentum Indicum* resembles one of the "mermaids" taxidermists fake from monkeys and fish tails, and his description of it is perfunctory.

Steller's sea ape bewilderment, which would have amused his shipmates, perhaps pushed him to take a new look at Bering Island's fauna, although his strong curiosity may have been enough to motivate him anyway. In any case, he began to record that fauna in an unprecedentedly systematic way. He not only collected and dissected it, standard activities for 1700s naturalists; he comprehensively observed its ethology and ecology, less familiar pursuits, not least in that they would not coalesce as scientific disciplines for over a century.

It was a more intact fauna than others were encountering in the 1740s. Linnaeus saw a Lapland long exploited by nomads and trappers, and the eastern North America that colonial naturalists like Catesby encountered had supported large farming populations for well over a millenium. Later naturalists exploring western North America and sub-Saharan Africa would see more diverse megafaunas, but Bering Island was exceptional even compared to those places. It was the closest thing outside Antarctica to an unknown land. At their brief Alaskan island landing, Steller had found a hearth of live coals surrounded by hastily abandoned artifacts, and they had encountered well-equipped natives several times during the return voyage along the Aleutians. But, although Asian and American cultural jetsam littered Bering Island's beaches (a poplar wood shutter and a fox trap made with *Dentalium* shells), Steller saw no evidence humans had ever been there. The animals, and not only the foolish sea cows, did not fear men.

"We had not yet reached the beach when a strange and disquieting sight greeted us," Steller marveled, "as from the land a number of sea otter came toward us in the sea, which from the distance some of us took for bears, others for wolverines." Ashore, arctic foxes swarmed everywhere, snapping at shoes and stealing "everything they could carry away, including articles that were of no use to them." The foxes indiscriminately tried to eat the dead and the sick and seemed innumerable. Steller wrote that he killed sixty with an axe in a day, but that it merely encouraged the others.

Faced with a wrecked ship, a dying commander, and a sick crew, Steller did not straightaway found the first zoological research station. He began by feeding sea otters to the men, saving the pelts to sell in the lucrative Chinese market if they ever got back to civilization. (He objected, however, when the men started killing just for pelts to gamble with.) Since the Bering Island otters often rested on shore, it was easy to corner and club them, and they killed between seven and nine hundred in nine months. The fresh meat helped cure scurvy, along with greens Steller gathered, and his ministrations finally gained him respect from officers and crew. But survival's brutalities (and its cruelties, as the men took revenge on foxes by imaginatively torturing them) did not dull his appreciation of the marvels he'd stumbled on.

"If they have the luck to escape," he wrote of the otters,

> they begin, as soon as they are in the water, to mock their pursuers in such a manner that one cannot look on without particular pleasure. Now they stand upright in the water like a man and jump up and down with the waves and sometimes hold the forefoot above the eyes as if they wanted to scrutinize you closely in the sun; now they throw themselves on their backs and rub their bellies and pudenda as do monkeys; then they toss the young ones in the air and catch them again, etc.
>
> Altogether a beautiful and pleasing animal, cunning and amusing in its habits, and at the same time ingratiating and amorous. They prefer to lie together in families, the male with its mate, the half-grown young and the very young sucklings all together. . . . Their love for their young is so intense that they expose themselves to the most manifest danger of death. When taken away from them, they cry bitterly, like a small child, and grieve so much that, as we have observed from rather authentic cases, after ten to fourteen days they grow as lean as a skeleton, become sick and feeble, and will not leave the shore.

Steller knew about the otters' love for their young partly because he and the men relished roast suckling sea otter, but he did not let his stomach interfere with his observations. The long winter of 1742 gave him plenty of time to study otters, a cast-up whale, a sea lion, and the fantastic but ubiquitous sea cows. "Every day for ten months during our ill-fated adventure I had a chance to watch from the door of my hut the behavior and habits of these creatures . . . ," he wrote of the latter. "[W]ith the rising tide they came in so close to shore that not only did I on many occasions prod them with a pole, but sometimes even stroked their backs with my hand."

His description included such an exact account of the sea cow's kelp diet that modern biologists have been able to determine the plants' genera:

> These gluttonous animals eat incessantly, and because of their enormous voracity keep their heads always under water with but slight concern for their life and security, so that one may pass in the very midst of them in a boat. . . . All they do while feeding is to lift the nostrils every four or five minutes out of the water, blowing out air and a little water with a noise like that of a horse snorting. While browsing they move slowly forward,

one foot after the other, and in this manner half swim, half walk like cattle grazing, half the body always out of the water. . . . Where they have been staying even for a single day, there may be seen immense heaps of roots and stems. Some of them when their bellies are full go to sleep lying on their backs, first moving some distance away from shore so as not to be left on dry land by the outgoing tide.

In spring, local otter decimation forced further exploration of the island, which led to the solution of a biological mystery. A party returned from the island's other side with the news that bull fur seals were arriving on the beaches there. "While on Kamchatka," Stejneger wrote, "Steller had already heard of these sea animals, which annually made their appearance in spring on the Kuril Islands, traveling north along the coast of Kamchatka, and congregating in the bay between Shipunski and Kronotski capes, only to disappear suddenly at the beginning of June." Steller had discovered where the fur seals went in summer, although the men were more interested in a new food source than pinniped life history. (When they butchered a bull, however, the putrid-smelling meat made them sick.)

"By the middle of June the great mass of breeding females had landed," Stejneger continued,

and Steller, who in his frequent trips to the south side . . . had watched the arrival of their incredible numbers, made it his particular object to study the doings of these remarkable animals. In a temporary hut, built for the purpose on a slight elevation in the middle of the rookery, sur-rounded by the noisy herds of fur-seals and families of sea lions, he spent six consecutive days carefully observing every detail of their habits of living—their matings, their fights, their parental care, the play and other doings of the new-born pups and their first attempts at swimming.

He not only observed, but with notebook in hand wrote down all he saw in elaborate detail with the result that the scientific world ten years later, when the St. Petersburg Academy published his great work, had a more accurate account of this hitherto unknown animal than of some of the most familiar species of Europe itself. In fact, when the American officials took possession of the Pribilof Islands [the largest rookery, in the Bering Sea off Alaska], and reported on the fur-seal herd acquired by the United States, their account contained but few additions and corrections to Steller.

Again, Steller did not proceed quite like a modern researcher. He and the crew were as careless of pinnipeds' lives as they were of otters' and sea cows', even more so, since bulls attacked fiercely when approached. "If I was asked to state how many I have seen on Bering Island," he wrote of the fur seal and sea lion rookery, "I can say without lying that it is impossible to make computation. . . . [They] covered the whole beach to such an extent that it was not possible to pass without danger to life and limb." He once spent hours on a knoll hemmed in by angry fur seal bulls. A common practice in such situations was to blind an offending bull by throwing rocks and watch his slow dismemberment by his rivals and the omnipresent foxes.

But Steller remained sympathetic. "The parents love their offspring exceedingly," he wrote of fur seals, which he perceptively called sea bears. He observed:

> The females, after parturition, lie in crowds upon the shore with their pups and spend much time sleeping. The pups, however, directly in the first days play together like children, and imitate their parents in playing at copulation, and practice at fighting until the one throws the other to the ground. When the father sees this, he rises up with a growl and hastens to separate the combatants, kisses the victor, licks him with his tongue, tries with his mouth to throw him upon the ground, and makes vigorous demonstrations of his love for the youngster, who struggles bravely against it. In short, he rejoices that he has a son worthy of himself.

He found his namesake sea lions more formidable: "These beasts are indeed terrible to look upon when alive, and they far surpass the sea bear in strength and size as well as in endurance of the different parts. They are hard to overcome and fight most viciously when cornered." Yet he observed them as intimately:

> They lay about me in every direction; they watched my fire and what I did, and did not run away any longer when I walked around them and took their pups and killed them and examined them. They practiced coition, fought jealously for their wives and for the best places, and fought most bitterly in just the same way and with the same motions and the same heat as the sea bears do.

He described sea lion life history elegantly:

> The reason why these beasts come hither in June, July, and August are for quiet, for parturition, for rearing and teaching the pups, and for copulation. Before and after this period they are found in great numbers on the

shores of Kamchatka. As to the food of these beasts, they prey upon fish and seal, especially, and also upon otter and other sea animals. The old ones eat little or nothing at all in June and July, but take their ease and sleep, and at the same time become very thin.

Steller finished his "great work," *De Bestiis marinis*, "Of Marine Beasts," on the island. It included drawings by an expedition artist of animals and dissections, and Steller also prepared skins and skeletons to take back as type specimens. Despite his protestations, the ship they used to escape, built from the remains of the *Saint Peter*, was too small for specimens except a sea cow palate. But he was able to dispatch the text and drawings to St. Petersburg in 1743 after the expedition's return to Kamchatka, although not without some characteristic fussing.

"But one must not suppose that these places do not contain more great and wonderful animals that are still unknown, beside those which I shall describe," he wrote in his introduction.

For if weather, time, and place had favored my desire I should have
enriched natural history with many curiosities of that sort, as indeed
I desired when I took the risk of this journey to parts so distant and
unexplored. Thus, for instance I describe the traces of a certain unknown
animal upon the island of Shumagin [one of the Aleutians the *Saint Peter*
visited], and I insert a sketch of a sea ape, and with this imperfect account
I must content myself and others.

Steller wrote quite a bit more on the voyage than "this imperfect account," not least his journal. He described several new seabirds, not only the giant, flightless Pallas's cormorant, which became a big part of the expedition menu, but "a special sea eagle with a white head and tail," probably the still extant Steller's sea eagle, and a "white sea raven . . . impossible to reach because it lights singly on cliffs facing the sea." The latter bird, like the sea ape, has never been identified or seen again.

Still, the frustration he expressed in his *De Bestiis* introduction proved justified. The drawings he had sent disappeared (including, maddeningly, the sea ape sketch), although copies of otter, fur seal, and sea cow ones survived. The text reached its destination, but Steller never benefited from his work. A new empress, Elizabeth, disliked science and foreigners, so he had no future in St. Petersburg, where his wife had sold his library as well as his furniture. He stayed in the Far East, telling Johann Gmelin that he had accomplished nothing of value during the Bering expedition, and blaming its officers for obstructing him. Rumor had it that he began drinking heavily, although he kept observing and writing. He taught school, then traveled north to collect plants, and again annoyed the authorities with complaints about mistreatment of natives. Accused of fomenting rebellion, he was on his way to exonerate himself in November 1746, when he died in the west Siberian town of Tyumen of a fever allegedly exacerbated by lying unconscious on his sledge while his drivers ate a leisurely meal in a roadhouse. Local louts rifled his grave for a red cloak he'd been buried in, and left his corpse naked in the snow until friends reburied it.

De Bestiis marinis received little attention even after its publication in 1751. "Struck down before the great Linnaeus had perfected the system which became the foundation stone of modern biology," Stejneger observed, "Steller's memory has become somewhat dimmed, and his achievements have chiefly served to glorify the names of other naturalists who many years later fell heir to his treasures of collection and observation." The work's Latin text has never been fully rendered in English, although the 1899 U.S. government publication on Pribilof fur seals Stejneger mentioned translated much of it.

In another way, Steller's discoveries were the reverse of obscure. As soon as the news of the Bering Island sea otters and fur seals hit the Kamchatka market, a wave of greed that had been building on the West Pacific shore swept eastward. When the ice melted in the spring of 1743, a ship sailed to Bering Island and returned with 1,600 otter skins and 2,000 fur seal skins. It was the beginning of a period that would nearly obviate Steller's achievement by obliterating his subjects.

NINETEEN

An Industrial Interlude

Steller described the Kamchatka fur industry in *De Bestiis*. Hunters killed up to forty sea otters an hour by hunting them with all means and at all times. They mobbed them in boats, forcing them to dive repeatedly: "From the rising bubbles they notice in which direction the animal goes. If the animal has a young one with it, this first loses its breath and drowns. . . . Finally the mother becomes so breathless and exhausted that it cannot stay under water for a minute. Then they dispatch it, either with an arrow or often, when near by, with the lance." The killing continued through the subarctic winter as hunters clubbed otters on the sea ice. "Generally such a storm and blizzard reigns that one can hardly keep one's feet, but nevertheless the hunters do not neglect the night time. They run along the ice without heed, even when it is drifting and being lifted on the waves to such an extent they now appear to be on a mountain and then to plunge into an abyss."

By similar relentless persecution, Russians exterminated Bering Island's otters in 1756 and moved on eastward along the Aleutians, looking for more. The first native Aleuts they encountered took a dim view of this and fought back, but the Russians were better armed. They soon discovered that they could get pelts more easily and pleasantly if they forced the skilled Aleut men to do the hunting in their skin *baidarkas* while they stayed on the islands with women and children as hostages.

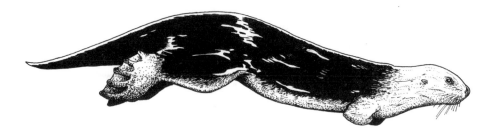

Figure 23. The sea otter, *Enhydra lutris,* was once abundant from Asia to Baja California, but the fur industry exploited it so relentlessly that it was mistakenly declared extinct in the early twentieth century.

Seventeen years after Steller's death, a native of Kodiak Island off Alaska described the industry's arrival:

> When we espied the ship at a distance, we thought it was an immense whale, and we were curious to have a better look at it. We went out to sea in our *baidarkas,* but soon discovered it was no whale, but an unknown monster of which we were afraid, and the smell of which made us sick. The people on the ship had buttons on their clothes, and at first we thought they must be cuttlefish, but when we saw them put fire into their mouth and blow out smoke we knew they must be devils.

Assured by Aleuts with the ship that they simply wanted to trade sea-otter skins for beads and other treasures, the Kodiaks brought out their furs. Then, "the Aleuts, who carried arms concealed about them, at a signal from the Russians fell upon our people, killing about thirty and taking away their skins. Those who attempted to escape in their *baidarkas* were overtaken by the Aleuts and killed."

Within three decades after Bering's voyage, the industry had seized the Aleutians and west Alaskan coast. "There are Russians settled upon all the principal islands between Oonalaska and Kamtschatka for the sole purpose of collecting furs," reported the narrator of Captain James Cook's last expedition, which explored from Oregon to the Arctic Ocean in 1778. "Their great object is the sea beaver or otter. I never heard them inquire of any other animal, though those whose skin are of inferior value, are also part of their cargoes." He thought the Russians sensible and businesslike, the Aleuts docile and industrious. Other natives, the Russians told him, were treacherous.

The Cook narrator did not note any great scarcity of marine mammals in the Aleutians, not having seen them before the decades of slaughter, but did remark on their abundance in other areas. Walruses, or "sea horses," as the English called them, were particularly numerous off the western Alaskan coast: "They lie, in herds of many hundreds upon the ice; huddled over one another like swine, and roar and bray very loud." The expedition provisioned itself with them: "[T]here were few on board that did not prefer them to our salt meat."

In the Nootka Sound area to the south, Cook's expedition found otters still common:

> The sea animals seen off the coast were whales, porpoises, and seals. Sea otters, which live mostly in the water, are found here. The fur of these animals, as mentioned in the Russian accounts, is certainly softer and finer than any others we know of; and therefore the discovery of this part of the continent of North America, where so valuable an article of commerce may be met with, cannot be a matter of indifference. . . . Mr. Coxe, on the authority of Mr. Pallas, informs us that old and middle aged sea otter skins are sold at Kiatcha by the Russians to the Chinese, for from 80 to 100 roubles a skin; that is, from 16 to 20 pounds each.

The narrator remarked that Northwest Indians, as yet unacquainted with the industry, "set no more value upon these than upon other skins . . . 'til our people set a higher price on them, and even after that, the natives . . . would sooner part with a dress made of these than with one made of the skins of wild cats or martins [sic]." The Northwest tribes learned the price of sea-otter skins soon enough, when the industry got to them. Having their own elaborate traditions of warfare and enslavement, however, they proved harder to subdue than the Aleuts. Increasingly under government control, the Russian industry grew more humane—toward humans.

Nine years after Cook's passage, when Captain Gerasim Pribilof found the islands in the Bering Sea off western Alaska that bear his name, the fur seal and otter populations there must have made Bering Island seem paltry. His ships returned from the voyage loaded with pelts, walrus ivory, and baleen, and returned the next year for more. The Russians exploited Pribilof otters so thoroughly that modern attempts to reestablish them there have floundered. If they had less success at destroying the fur seal rookery, it was not for lack of trying. There were simply too many to exterminate in a few decades. An estimated 5 million northern fur seals, *Callorhinus ursinus,* originally bred on the islands ("estimated" is the operative word—nobody really knows the number), and the Russians, in the next fifty years, only managed to reduce that to an estimated 3 million.

But help was on the way. Following Cook, British and American whalers moved into the Pacific. Although they favored killing sperm, right, and bowhead whales for their oil and baleen, other resources would do, and they were glad to diversify into furs. Sea otters remained the item of choice as long as they lasted. They were common enough in San Francisco Bay when the Spanish arrived in 1776 that boatmen could club them with oars.

"It is astonishing beyond measure," wrote Jean-François de Galaup La Pérouse, who led a French expedition down the coast in 1786, "that the Spaniards, having so close and so frequent contact with China by way of Manila, should have been ignorant up to now of the value of this precious skin. . . . Before this year, an otter skin has been worth no more than two rabbit skins." Since France was unlikely to get in on the trade, La Pérouse thought competition with the Russians would benefit Spain, "for it is in the nature of exclusive privilege to bring death, or at least torpor, into all the branches of commerce and industry, and it is the part of liberty to give them all the activity of which they are capable." Within a decade, hunters were killing hundreds of otters a week in San Francisco Bay. In 1811, Aleuts working for three American ships headquartered in Drake's Bay took an estimated 10,000 otter skins along the California coast.

Another luxury item, fur seals, bred in large numbers on California islands. Biologists are unsure now whether most of those were northern fur seals or a southern species, the Guadalupe fur seal, *Arctocephalus townsendi.* Hunters did not discriminate. According to one source, British whalers killed more than 8,000 fur seals on the San Benito Islands off Baja in 1805, and Americans killed over 70,000 on the Farallones in 1810. According to another, whalers killed 130,000 on the Farallones in 1808 and 1809.

The Russians reached the California coast at the same time as the Americans and British and, being better organized, continued to dominate the fur industry despite La Pérouse's hopes for liberal competition. Fort Ross on the Mendocino coast became their headquarters. "Their skill in killing otters seems incredible," wrote a Californian of the Aleut hunters. "Whenever two of them jointly pursue an otter, it will not escape, because they quickly tire the animals, get near them, and strike them." A Russian station on the Farallones superseded the whalers in killing fur seals and also exploited seabirds. By the 1820s, the Russians were killing 50,000 western gulls a year on the Farallones alone.

When a Russian captain, Otto von Kotzebue, led an expedition down the coast in 1824, the otter supply was dwindling. "[S]o scarce are these animals now become even here," he wrote of southeast Alaska, "that the numbers caught only suffice to cover the expenses of maintaining a force sufficient for protection against the savages." Things were better in California, where otters, "though at present scarce even here," were still numerous south from Fort Ross. Kotzebue foresaw a quick end, however: "Its value advances yearly, with the increasing scarceness of the animal; it will soon entirely disappear, and exist only in description to decorate our zoological works."

When the Russians sold off their California possessions and withdrew north in 1841, it seemed a shrewd move. They had made several times as much from furs as they would from selling Alaska to the United States in 1867. And by then they could no longer control the exploitation of northern fur seals on the Pribilofs, because fleets of British and American schooners were pursuing the seals on their pelagic fishing grounds, shooting pregnant females along with everything else, and failing to recover the large percentage of dead animals that sank.

The American takeover on the west coast renewed the slaughter by removing restrictions that had developed within the Spanish and Russian systems. Freelance hunters shot otters and Guadalupe fur seals from boats or the shore, wasting much of their kill, while larger concerns went after the oil, baleen, and leather markets. When big pelagic species became scarce, they attacked the smaller, coast-hugging gray whales, then the smaller elephant seals, then the smaller sea lions, pursuing them to mating lagoons and island rookeries. A little later, the Alaskan walruses had their turn, shot for their hides and fat. Seabirds were not neglected. An estimated 400,000 common murres nested on the Farallones originally. In the 1850s and 1860s, some 12 million murre eggs graced the Bay Area's tables, and by 1910, around 20,000 murres nested on the Farallones.

The American industry did not produce only breakfasts, corsets, coats, fuel, and lubricants. It also engendered a kind of mirror image of Georg Wilhelm Steller. In 1874, an American named Charles Melville Scammon published a book entitled *Marine Mammals of the North Western Coast of North America,* which, with Steller's *De Bestiis marinis,* comprised most of the pre-1900 contribution to understanding of west coast sea mammals. But the two authors' careers were opposite. Steller was a European Enlightenment naturalist who accidentally helped to launch the industrial slaughter. Scammon was an American Gilded Age seaman who participated in the slaughter, then, as an afterthought, tried to mitigate it.

Born in Maine in 1825, Scammon went to sea at age seventeen and got his first command in 1848. The next year, he captained a merchant ship to California. Apparently, the marine mammal industry had not attracted him, and he took it up only when, in the backwash of the gold rush, west coast commands became scarce. But he adapted well, seeing not the grubby scramble Steller described, but businesslike sport. He detailed elephant seal slaughter on an 1852 voyage to Baja California with a craftsman's satisfaction:

> The mode of capturing them is thus: the sailors get between the herd and the water; then, raising all possible noise by shouting, and at the same time flourishing club, guns, and lances, the party advance slowly toward the rookery, when the animals will retreat, appearing in a state of great alarm. Occasionally an overgrown male will give battle, or attempt to escape; but a musket-ball through the brain dispatches it; or someone checks its progress by thrusting a lance into the roof of its mouth, which causes it to settle on its haunches, when the men with heavy oaken clubs give the creature repeated blows upon the head, until it is stunned or killed. After securing those that are disposed to show resistance, the party rush on to the main body. The onslaught creates such a panic among these peculiar creatures, that, losing all control of their actions, they climb, roll, and tumble over each other. . . . The quantity of blood in this species of the seal tribe is supposed to double that contained in an ox, in proportion to its size. . . . After the capture, the flaying begins.

The rendered blubber, Scammon noted, was "superior to whale oil for lubricating purposes." He captained such voyages for many years and, later in the 1850s, mirrored Steller by discovering another unknown world—the gray whale calving lagoons in southern Baja. "This species of whale manifests the greatest affection for its young," he wrote, "and seeks the sheltered estuaries lying under a tropical sun as if to warm its offspring into activity and provide comfort, until grown to the size nature demands for its northern visit." As the industry rushed in after him, the whales' subtropical refuge dissolved even faster than had the sea cows' subarctic one.

Scammon saw a kind of poetry in the killing. "Here the objects of pursuit were found in large numbers," he wrote of a Baja lagoon,

> and here the scene of slaughter was exceedingly picturesque and unusually exciting, especially in a calm morning, when the mirage would transform not only the boats and crews into fantastic imagery, but the whales, as

they sent forth their towering spouts of aqueous vapor, frequently tinted with blood, would appear greatly distorted. At one time, the upper sections of the boats, with their crews, would be seen gliding over the molten-looking surface of the water with a portion of the colossal form of the whale appearing for an instant, like a spectre in the advance; or both boats and whales would assume ever-changing forms, while the report of the boat-guns would sound like a sudden discharge of musketry.

Sometimes he found it idyllic:

[F]or many years, the Sea Otter hunters along the coasts of California and Oregon were made up of nearly all the maritime nations of Europe and America . . . hardy spirits who preferred a wild life and adventurous pursuits rather than civilized employment. The distance coasted in their lightly constructed boats, the stealthy search for the game, and, when discovered, the sharp-shooting chase, gave these hunting expeditions a pleasant tinge of venture. . . . From day to day, if the weather is pleasant, they cruise in search of the animals, landing to pass the night in places well known to them. . . . Their evening meal . . . is partaken of with hearty relish; then follow the pipes, which are enjoyed as only these men of free and easy life can enjoy them. Relieved from all care, these adventurers talk of past exploits or frolics, and finally roll themselves in their blankets for an invigorating sleep in the open air.

The hunt did have disturbing moments for Scammon, like one raid on a California sea lion rookery:

The herd at this time numbered seventy-five, which were soon dispatched by shooting the largest ones, and clubbing and lancing the others, save one young Sea Lion, which was spared to ascertain whether it would make any resistance by being driven over the hills beyond. The poor creature only moved along through the prickly pears that covered the ground, when compelled by his cruel pursuers; and at last with an imploring look and writhing in pain, it held out its fin-like arms, which were pierced with thorns, in such a manner as to touch the sympathy of the barbarous sealers, who instantly put the sufferer out of its misery by a stroke of a heavy club.

And more than the occasional twinge of compassion interrupted Scammon's commercial sportsmanship. Although he liked to act the bluff Yankee skipper, he was not content just to navigate and then look on as his crews skillfully slaughtered, flayed, and flensed. Curiosity beset him. Like Steller, he measured dimensions, examined organs, made drawings, and kept a detailed journal. This was not "free and easy," but hard work in addition to his navigation duties: "It is hardly necessary to say that any person taking up the study of marine mammals, and especially the Cetaceans, enters a difficult field of research, since the opportunities for observing the habits of these animals under favorable conditions are rare and brief. My own experience has proved close observation for months, and even years, may be required before any single new fact in regard to their habits can be obtained."

Scammon may have liked the industry less than he let on. He quit in 1863, and during the rest of the Civil War commanded U.S. Revenue Marine ships patrolling the west coast. Then he went to work for Western Union scouting a Bering Strait route for a transoceanic telegraph cable. The project failed, but not before he had spent two years in the remote North Pacific, coming into contact with scientists like the Smithsonian's William Dall. The experience encouraged him to write about his observations. As with Steller, this had mixed results. It brought him some notice, but, because of distance from scientific centers, it also led to some appropriation of his material.

Edward Cope was prominent in this, not surprisingly, since Scammon's writing debut coincided with his youthful cetacean fling. In 1869, Cope got a copy of an unpublished Scammon article on west coast whales from Joseph Henry, a Smithsonian naturalist, and attached an edited, sometimes inaccurate, version of it to a synopsis of whale classification he published in the Philadelphia Academy of Natural Sciences' *Proceedings.* He justified himself on the grounds that Henry had referred the piece to him "with the request to publish such parts as could be deemed valuable to zoology." He hadn't asked Scammon's permission, however, and he also used some of the material, unacknowledged, in his own synopsis. Scammon complained to the Smithsonian's influential Spencer Baird, who knew of young Edward's bumptiousness and promised to chastise him. It probably encouraged Cope's growing caution in marine mammal publications.

Cope's behavior was not entirely self-seeking. He praised Scammon's work, calling it "the result of many years' observation in an unexplored and with difficulty explorable department of zoology." He named a pilot whale, *Globiocephalus scammonii,* "in honor of Captain Scammon, who has furnished us with a mass of information on the Marine Mammalia, and an amount of novelty connected with it seldom equalled in the history of zoology." Scammon drew on Cope's work, too. Many of the cetacean names he used were Cope's, including those of gray, finback, blue, humpback, and pilot whales, and of various dolphins. Some of his ideas may reflect Cope's evolutionary speculations, as when he wrote that "eminent naturalists" thought walruses might be "the connecting link between the mammals of the land and those of the water."

In any case, unlike Steller, Scammon was able to to benefit from his work. In 1868 he became a contributor, along with Samuel Clemens, John Muir, Joaquin Miller, and other celebrities, to San Francisco's popular *Overland Monthly*, with articles that ranged along the coast from Baja to Alaska. The first, on fur seals, came out in 1869. The articles formed the core of his *Marine Mammals*.

Scammon's writing had greater scope than Steller's—he described the whole west coast, and every marine mammal then known. This led, despite his lack of training, to scientific reflections beyond Steller's. Although there was little west coast marine fossil record then, as Cope's fumblings with his San Diego "*Eschrichtius*" showed, Scammon touched on evolutionary relationships. Sometimes he did so at second hand, as with walruses, sometimes from personal observation, as when he speculated that the sperm whale might have "a higher organization than any other species of Cetacea. Its massive form . . . composed of bone, flesh, and sinew which has a finer texture than that of the rorqual or the mysticetes."

Scammon sensed the behavioral complexity that came to light a century later. He suspected sperm whales might have "the faculty of communicating with each other in times of danger, when miles (and some observers say leagues) distant." He thought gray whales "possessed of unusual sagacity," as shown by the "numerous contests with them" in which "enraged animals have given chase to the boats which only found security by escaping to shoal water or to shore." He noted the "amorous antics" of humpbacks. "At such times," he wrote, "their caresses are of the most amusing and novel character." He thought harbor seals "endowed with no little sagacity," and despite their apparent muteness compared to sea lions, noted times when many met "in the neighborhood of rocks or reefs distant from the main land" and became "quite playful . . . leaping out of the water and circling around upon the surface."

Beneath their differences, the two men had much in common. Curiosity often led Scammon to a sense of wonder like Steller's. "The sea was quite smooth, and not a breath of wind was stirring," he wrote of a Dall's porpoise sighting, continuing:

At first we could hear a harsh rustling sound, as if a heavy squall of wind accompanied by hail, was sweeping over the otherwise tranquil sea; and as the moon burst through the clouded sky, we could see a sheet of foam and spray sweeping toward us. In a few moments the vessel was surrounded by myriads of these Common Porpoises, which, in their playful movements, for the space of one hour, whitened the sea all around us as far as the eye could discern, when they almost instantly disappeared.

Scammon wondered not only at spectacular creatures, but at quiet ones, like harbor porpoises. As he wrote, they were never seen in large groups, and seldom jumping, "their general habit being to make a quick puff and turn as soon as they appear above water, apparently choosing the darkness below rather than the light above."

Scammon's wonder expanded occasionally into Steller's mythic awe. His scrapbook contained a news article about a thirty-foot-long "wonderful fish," purportedly seen by Spencer Baird, that had fins, a sharklike tail, and two huge legs ending in webbed feet. Despite the more settled times, he once saw something on the California coast almost as surprising as a sea cow. Referring to the "banded seal" (now the ribbon seal, *Histriophoca fasciata*), a "beautifully marked" but "very little known" species of the Bering Sea, he wrote: "In April 1852 we observed a herd of seals upon the beaches of Point Reyes, California; these, without close examination answered to the description given by Gill. . . . Although we were quite confident that the seals . . . were the same. . . . still it is a remarkable fact that we have never seen this species on the California coast again." As Steller saw the end of 30 million years of North Pacific sirenian evolution, Scammon may have glimpsed the last of a pinniped's ice age migratory cycle.

Scammon's story lacks the high drama of Steller's. He didn't discover an unknown species, never again seen by naturalists, and he didn't have a tragic end. A devoted wife bore him three sons, and he died in bed at eighty-six. Still, a cloud seems to have settled over his life after 1870, as though he too had gone too far into remote realms to return to society. His personality, prickly like Steller's, and exacerbated by long shipboard tensions, contributed to this. Just before *Marine Mammals*' publication, he alienated William Dall, his main scientific connection, apparently by accusing him too of appropriating material.

Despite good notices, *Marine Mammals* sold only half of a thousand-copy first edition. The rest burned in the San Francisco earthquake, and the book remained out of print until nascent environmentalism resurrected it in 1968. Bad health forced Scammon to retire from the sea for many years soon after the book emerged, and although he went on expeditions later in life, he may have felt that his career had fallen short. He spent his last years in Oakland, seldom speaking, writing things that seem to have disappeared.

Scammon suffered one misfortune that Steller was spared. Like other Enlightenment naturalists, the German was not excessively concerned about the future of the animals he studied. He was more concerned than most, since he liked the animals and disliked the greedy mass slaughter already underway. But his optimism about sea cows supplying Kamchatka suggests that he did not clearly foresee the holocaust to come. He was religious and may have shared the prevalent pre-Cuvierian view that extinction was impossible. Again, Scammon's outlook was the reverse. Like Steller a man of his time, he was agnostic and pessimistic under his businesslike exterior, with a free-floating sense of guilt. Extinction was a present concern to him, almost an obsession judging from its frequent mention in *Marine Mammals.*

The gray whale's coastal migration, he wrote, exposed it to persecution from virtually every society along the way, from "savage tribes" to "civilized whalemen . . . thus hastening the time of its entire annihilation." Elephant seals, "owing to the continual pursuit of the animals," had become "nearly if not quite extinct on the California coast, or the few remaining have fled to some unknown point of security." Even sea lions would "soon be exterminated by the deadly shot of the rifle, or driven away to less accessible haunts." By the time Scammon got to sea otters, he seems to have tired of forecasting doom, merely asking plaintively if "these sagacious animals" might not have fled "from those places on the coast of California where they were so constantly pursued to some isolated haunt" where they could "remain unmolested."

Scammon saw poetry in extinction as well as in killing:

The civilized whaler seeks the hunted animal further seaward, as from year to year it learns to shun the fatal shore. None of the species are so constantly and variously pursued as the one we have endeavored to describe; and the large bays and lagoons, where the animals once congregated, brought forth and nurtured their young, are already early deserted. The mammoth bones of the California Gray lie bleaching on the shores of the silvery waters, and are scattered along the broken coasts, from Siberia to the Gulf of Alaska; and ere long it may be questioned whether the mammal will not be numbered among the extinct species.

Bones were less picturesque than live whales, however, and even they vanished soon enough, collected, as Cope noted in 1879, for fertilizer. When he died in 1911, Scammon probably thought many species had gone before him. Northern elephant seals were considered extinct in 1870. Guadalupe fur seals, having been declared extinct in 1840, were declared so again in 1894. California gray whales also underwent multiple extinction, first around 1870, then around 1910. An exhaustive search for sea otters in 1925 found none, probably a good thing, since scientists might have killed survivors for museum specimens, as they killed some elephant seals found on Guadalupe Island in 1892, apparently exterminating them again. Walruses became so rare that native tribes who depended on them for subsistence starved. Pribilof Islands fur seals numbered between 300,000 and 130,000 by 1911.

Hardly anybody seemed to care much. Unlike the bison slaughter that had appalled eastern journalists and reformers, the marine holocaust happened so far away that few noticed. When John Muir traveled through southeast Alaska during summers from 1879 to 1881, he remarked on salmon abundance but not on sea otter absence, not surprisingly, since otters had been gone for a century. In his *Travels in Alaska,* he mentioned seeing unspecified whales and porpoises twice, unspecified seals once: "The whiskered faces of seals dotted the open spaces between the bergs, and I could not prevent Johnny and Charley and Kachadan [his Tlingit guides] from shooting at them. Fortunately, few, if any, were hurt."

When they explored the southern coast sixty years later, John Steinbeck and Ed Ricketts found a similar marine tetrapod scarcity and described it as laconically. They mentioned pelicans and a "tawny, crusty" sea lion off Monterey, the pointless harpooning and slow death of a hawksbill turtle off Baja's west coast, and sleeping seals and a "school of whales," both of unknown species, in the Sea of Cortez. One of the whales came close enough to wet the boat when it blew, but this did not please them: "There is nothing so evil-smelling as a whale, anyway, and their breath is frightfully sickening. It smells of complete decay. . . . We felt deeply the loneliness of this sea."

Nudged by scattered voices like Scammon's, governments began sluggish attempts to regulate hunting by the turn of the nineteenth century. But an

1897 pronouncement by a prominent American scientist, Professor T. C. Mendenhall, manifested a common attitude among the few with much awareness of the slaughter:

> Two considerations call for the protection of Alaska seals, the sentimental and commercial. The former may be dismissed, as it has been in cases of far more intimate contact between man and the species exterminated. The commercial consideration is one that ought not to be difficult to deal with. . . . The practical extinction of the herd in the near future seems assured. The United States may have the pick of what remains by whole-sale killing on the islands.

TWENTY

Intimations of Communication

The west coast's miraculous multiple marine extinctions show how much our sense of reality depends on received opinion. An excellent Point Reyes field guide published in 1988 described river otters as "apparently extinct" there because of "a combination of trapping and degradation of streams." I saw river otters in a pond at the National Seashore twice in the 1990s, but I'd have hesitated to insist if someone had doubted me. On the other hand, although the field guide also called black bears "probably extinct," I didn't doubt the existence of one that raided the dumpster of the Limantour Youth Hostel near Point Resistance in 2003, because, although I didn't see it, newspapers reported it.

I didn't report the river otters for lack of proof. People who didn't mention seeing sea mammals after their various extinctions probably had a similar motive, along with fear that reported animals might be killed. Chatting in the early 1930s with a Big Sur woman, Ed Ricketts "was astonished to hear her describe animals living in the surf which could only be sea-otters, since she described accurately animals she couldn't have known about except by observation." Ricketts reported this to a "learned institution" but got no response. (Apparently, some biologists had known about the Big Sur otters since at least 1915 but had kept quiet.) "It was only when a reporter on one of our more disreputable newspapers photographed the animals that the public was informed."

Anyway, all the multiply extinct mammals except Steller's sea cow eventually resurfaced. A sea cow was reported from Bering Island in the 1850s, but Leonhard Stejneger, who spent over a year there fairly soon afterward, deemed it a female narwhal. Sea cows had been rumored elsewhere, but Stejneger thought *Hydrodamalis* had been confined to Bering Island in historical times, and he considered Steller's impression of its numbers there exaggerated:

> I should regard fifteen hundred as rather above than below the probable
> number. It must be remembered that the sea cow was an extremely bulky
> animal, twenty-four to thirty feet long, which lived chiefly near the mouth
> of rivulets, feeding on sea-weeds, especially the large Lamellarias. There
> are hardly more than fifteen places on the island which could afford them
> suitable grazing grounds, and if each of these were regularly visited by an
> average of one hundred animals, one would easily be impressed by their
> numbers, especially if divided up into five or ten herds of from ten to
> twenty individuals.

Fortunately, other species were less localized. A few years after 1900, some elephant seals that had survived the 1892 scientific expedition showed up at Guadalupe Island. In 1931, the headman of an Aleut village on Amchitka Island took a biologist to a secluded cove there to show him the first officially recorded sea otters in many decades—a mother and pup. In 1946, some gray whales turned up, appropriately, since the International Whaling Commission (IWC) was established that year. Although Guadalupe fur seals had been declared extinct again in 1928 and 1949, scientists found fourteen in a sea cave on that island in 1954.

The four species have not dropped out of sight again, a circumstance to which things like the IWC and the 1972 Marine Mammal Protection Act undoubtedly contributed. Even though the past's huge profits were gone, sealing and whaling continued right up to (and after) the enforcement of such protective measures. Two stations that opened in San Francisco Bay in 1956 killed 3,400 whales of eight species in the next thirteen years. But softening markets for marine tetrapod products probably had as much to do with their survival. By the mid-twentieth century, most species literally had outlived their industrial usefulness, replaced by chemicals, and could proceed with rebuilding their populations if they did not run afoul of industries replacing those that had decimated them.

The multiply extinct species, as if in compensation, have not done too badly, although, being descended from a few individuals, they have lost much of their genetic diversity, which may affect their future survival. Eastern Pacific gray whales and northern elephant seals may be regaining their historical numbers and ranges on the west coast. Gray whales probably number over 20,000, moving between Baja and the Bering and Chukchi seas, and the elephant seal population may soon reach 200,000, with rookeries from Guadalupe Island off Baja to Point Reyes. (First occupied in the 1980s, the Point Reyes rookery was colonized by overflow from the Farallones, colonized in the 1960s, and from Año Nuevo, colonized in the 1950s.)

Sea otters have regained parts of their original range and rebuilt their population to perhaps half of an estimated original size of 300,000. But some reintroduction attempts like the one in southern Oregon have failed, and their numbers have suffered from nuclear tests, oil spills, diseases, human persecution, and other factors. Another important factor may be that, when otters are removed, sea urchins can become so abundant that they destroy giant kelp forest by overgrazing young plants, thus destroying otter habitat.

The waters off Point Reyes probably had bigger kelp beds in the past, for example, since the continental shelf there is wider than anywhere else on the coast. The Gulf of the Farallones was known for large otter populations. But the kelp beds visible from Point Resistance now seem scanty compared to the ones off Big Sur, where otters still live. (Even sea urchins seem scanty in Drake's Bay; at least I found none in rock pools exposed at one extreme minus tide.) Otters seem to have trouble recolonizing such areas, and very few have visited Point Reyes since hunters killed the last recorded native ones in 1848. Even where otters seem well established, as in central California, their numbers continue to fluctuate dangerously. In southwest Alaska, long an otter stronghold, the species declined so sharply after 1980 that it was designated as threatened in 2005.

Guadalupe fur seals are the least understood of the multiply extinct species. Originally, it was assumed that sealers had killed large numbers on the Farallones, but those may have been northern fur seals, so the species may never have been common north of the Channel Islands. In any case, its recovery has been the slowest of the four species. A small population lives off Baja, and vagrants are occasionally sighted at the Channel Islands.

Two west coast marine mammal species have come off relatively lightly. Although hunted by early sealers, the California sea lion never disappeared and now numbers well over 100,000 in the eastern Pacific, although a western Pacific population may be extinct. It breeds from the Sea of Cortez to the Channel Islands and migrates as far north as Alaska in summer. Industry never seriously pursued noncolonial harbor seals, although fishing interests persecuted them, sometimes with dynamite, and human presence drove them from beaches, particularly in southern California. An estimated 300,000 live from Alaska to Baja now, and they have reclaimed southern California to the extent that CBS News, in July 2004, reported that they were monopolizing a beach at La Jolla near San Diego. ("You can't have a conversation with them," an official complained. "They don't speak English.") But even harbor seals aren't immune to the extinction plague. Populations in the Gulf of Alaska and Prince William Sound underwent declines of up to 90 percent in the 1970s.

Other marine mammal populations are less healthy. Although the northern fur seal population rose to over 2 million in the years after a 1911 treaty outlawed pelagic sealing, it began falling again after 1948. For unclear reasons, it has been dropping by 6 percent a year since 1998, and less than a million now breed on the Pribilofs. (Northern fur seals began breeding on the Farallones again in 1996 and produced 80 pups in 2006, but that hardly makes up for the decline.) Steller's sea lions have declined even more drastically. When I first came to the west coast in the late 1960s, the species seemed as common as the California one, and, in fact, there were over twice as many Steller's, although most lived north of California. Since then, again for unclear reasons, although industrial preemption of fish stocks probably has contributed, the Steller's population has plunged well below that of the California sea lion.

Harbor porpoises are thought to be declining, for unclear reasons, and orcas and Dall's porpoises may be, too. The humpback whale, the only commonly seen west coast mysticete besides the gray, once had a North Pacific population estimated at 15,000, but, again for unclear reasons, it has not recovered to anywhere near that number. There aren't 15,000 humpbacks in the world today. Most cetacean species are so uncommon, wide-ranging, or elusive that their west coast status is vague. The IWC's 1986 moratorium on commercial whaling protects them from wholesale slaughter for the present, if not from pirate whaling and threats like drift nets and water pollution, but Japan, Iceland, Norway, and other prowhaling nations keep trying to rescind it. They nearly succeeded in 2006.

Declines may be interlinked in bewildering ways. According to one theory, the sea otter's disappearance from Bering Island by 1756 contributed to the sea cow's extinction a decade later as proliferating sea urchins, formerly otter prey, devoured the sirenians' kelp pastures. According to another theory, reduction of North Pacific and Bering Sea whale populations by 86 percent in the late 1960s (factory ships killed a half million large whales from 1949 to 1969) forced orcas to prey more heavily on other marine mammals, causing the western Alaskan declines, first of fur seals, then of harbor seals, then of Steller's sea lions, then of sea otters. There is good evidence that orca predation has reduced sea otter populations in the area. Orcas have been seen eating otters in significant numbers, and otters in orca-inaccessible areas have not declined.

Such theories have their problems. The evidence regarding the sea cow's Bering Island demise has long disappeared, and sea otter disappearance cannot explain sea cow extinction everywhere else, since otters outlived sea cows elsewhere. Evidence of prewhaling orca predation on large whales is also scarce. When North Pacific whales were numerous, Charles Scammon saw orcas attacking even the largest, but thought they fed more on dolphins and porpoises. Still, no better explanation than orca predation is known for the pinnipeds' and otters' recent declines. Scammon described orcas circling sea lion rookeries, greedily gulping inexperienced pups.

If commercial factors like the fur and whaling industries can have disastrous long-range effects on so many nontarget species, it is ominous for coastal ecosystems overall. And marine tetrapods generally have undergone demographic roller-coaster rides in the past half century. West coast brown pelicans faced extinction in the 1950s and 1960s because of DDT contamination of their nesting waters. Oil spills, climatic fluctuations, and gill netting caused a drastic decline in California common murre populations during the 1980s. In the 2000s, abnormally weak seasonal upwellings have reduced food for murres, pelicans, and most other fish-eating tetrapods. Leatherback turtles face extinction now because of drift netting on the high seas and exploitation of nesting beaches.

Still, the survival so far, however precarious, of all these industrially obsolescent organisms undercuts Professor Mendenhall's 1897 dismissal of sentimental considerations. If sentiment's futility was self-evident to him, it was less so to people like Steller and Scammon, whose activities were one reason why fur seals aren't extinct as Mendenhall predicted. Dismissing those considerations gets harder as knowledge grows and the boundaries between sentiment and realism become increasingly porous. Ecology suggests that living marine tetrapods can be as valuable as dead ones. (Kelp from healthy sea otter habitat, for example, can produce more income than pelts or shellfish.) Ethology hints that some may be "rational beings."

Steller's sea ape seems prophetic of the boundaries' porosity, with its "strange habits, quick movements, and playfulness"—raising itself out of the water "up to a third of its length, like a human being," doing "juggling tricks" with seaweed, watching the *Saint Peter* and its crew "with admiration" for two hours. So does a story Scammon repeated about a blue whale that followed a California-bound ship for twenty-four days:

> A week ago, we passed several, and during the afternoon it was discovered that one of them continued to follow us and was becoming more familiar, only coming up to breathe. A great deal of uneasiness was felt, lest in his careless gambols he might unship the rudder or do us some other damage. It was said that the bilge-water would drive him off, and the pumps were started, but to no purpose. At length, more violent means were resorted to; volley after volley of rifle shots were fired into him, billets of wood, bottles etc. were thrown upon his head with such force as to separate its integument; to all this he paid not the slightest attention, and he still continued to swim under us, keeping our exact rate of speed, whether in calm or storm, and rising to blow almost into the cabin windows. He seemed determined to stay with us until he could find better company. His length is about eighty feet; his tail measures about twelve feet across; and in the cabin as we look down into the transparent water we can see him in all his huge proportions.

Such intimations of communication went unexplored until the mid-twentieth century, for the same reason that marine tetrapods' evolutionary past largely did—lack of evidence. When exploration began, a figure similar in some respects to the fossil hunter Douglas Emlong was influential. As Emlong's fossils helped to illuminate a largely unknown past, John Lilly's dolphin research helped to illuminate an unknown mentality.

Like Steller and Scammon, Emlong and Lilly were opposites in many ways. Emlong was an impoverished introvert with whom, as Clayton Ray put it, "people generally and wisely preferred only superficial interaction sufficient to utilize his collections." Lilly was a wealthy extrovert who attracted many friends and followers. The heir of a Minneapolis banking fortune, he had successful mainstream careers as a biologist and psychoanalyst before he began to study dolphin intelligence in the 1950s. In the 1960s and 1970s, his dolphin work reached a wide audience and had a major, if mixed, influence on human-cetacean relations.

But both men shared a certain eccentric brilliance. Both were nonprofessionals in the marine tetrapod field, and both entered it for reasons as much aesthetic as scientific. Emlong's poverty and Lilly's wealth had the convergent effect of allowing them to pursue their agendas outside constraints normally affecting researchers. Although Lilly's money let him live longer than Emlong, from 1915 to 2001, the periods they spent exploring marine tetrapod paleontology and psychology were roughly equivalent. Even Lilly's wealth was not enough to realize his goals, and both men veered into New Age pseudoscience after financial and temperamental obstacles stymied their explorations.

Lilly first approached cetacean psychology simply as a way to study the mind-brain relationship. "When John began to be interested in dolphins, they represented little more to him than the opportunity to explore large brains," a biographer wrote. "He might experiment on these creatures using the techniques he could not use safely on himself or ethically on other human subjects." Large dolphin brains, Lilly thought, might reveal more about the mind than the small rat or cat brains most psychologists were studying.

As he heard stories about cetaceans, however, Lilly sensed another possibility—exploring a nonhuman intelligence. He heard about an orca population, for example, that had begun to avoid whaling boats with harpoon guns after such a boat accidentally shot one: "Yet the orcas in the area continued to cruise and feed around boats without harpoons. Apparently the orcas were able to distinguish boats with harpoon guns in the prows from otherwise identical boats without harpoon guns. Few of the other orcas had actually witnessed the shooting. The injured orca, or perhaps another who saw the shooting, must have communicated complex information, including a description of the dangerous boats, to hundreds of others."

Since orcas are giant dolphins, their behavior made Lilly suspect that he might study smaller species not only by putting them through experimental learning situations, as with rats in mazes, but by communicating with them. If dolphins communicated with each other in complex and sapient ways, perhaps they could do so with humans. The best way to do this, he saw, would be to learn dolphin "language," to translate the diversity of sounds they made and learn to mimic them. But the obstacles to this are enormous, since dolphins communicate underwater at much higher frequencies than humans. So he tried the next best thing. He started a lab in the Virgin Islands where he tried to teach bottlenose dolphins human language.

According to Lilly, learning human language was as much the dolphins' idea as his own. Cetaceans were thought incapable of mimicking words

Figure 24. *Orcinus orca,* literally "the whale from hell," was originally called the whale killer because it can prey on large whales. Highly intelligent, it may have sophisticated communication skills.

when he started, but his "working hypothesis of advanced capability" made him "listen to some rather queer noises the dolphins were making" and suspect that they were attempts at imitating human speech: "We began to have feelings which I believe are best described by the word 'weirdness.' The feeling . . . came on us as the sounds of this small whale seemed more and more to be forming words in our own language. We felt we were in the presence of Something, or Someone, who was on the other side of a transparent barrier which up to this point we hadn't even seen."

The lab had some success. Dolphins mimicked human words with their blowholes and used them in contexts that researchers could understand. The communication was limited, however. Lilly did not gain that much insight into how dolphins perceive reality and share their perceptions with each other. And dolphins died during the research, some apparently deliberately by refusing to eat. Lilly freed the survivors and closed the lab in 1967.

The dolphin lab's baffling mixture of success and failure evidently pushed Lilly past experimental empiricism. Convinced that large-brained cetaceans have high intelligence, he speculated that their minds might have superhuman capacities that could mimic high technology:

> "The theory is as follows: The sperm whale's brain is so large that he
> needs only a fraction of it for use in computations for his survival. He
> uses the rest of it for functions about which we can only guess. . . . If
> a sperm whale, for example, wants to see-hear-feel any past experience,

his huge computer can reprogram it and run it off again. His huge computer gives him a reliving, as if with a three-dimensional sound-color-taste-emotion re-experiencing motion picture. He can thus review the experience as it really happened. He can imagine changing it to do a better job next time he encounters such an experience.

Lilly also speculated that toothed whales' echolocation might give them the power to "see" inside the bodies of other organisms, thus providing insights into others' emotions bordering on telepathy. His books about such ideas were bestsellers and made the concept of cetacean superconsciousness popular with the public. They contributed to the vogue that boosted whale-watching participation in California alone from a few hundred a year in the late 1950s to over a million in the late 1990s. They were less popular with marine biologists, who saw little evidence for his speculations. Lilly didn't mind that, although he was upset when the United States Navy adapted his experimental dolphin-teaching techniques to military purposes.

"If we 'teach' them to aid our underwater work in the sea as glorified 'seeing eye dogs' of the Navy, of oceanographers and divers," he wrote, "we are far from my goals for them." He dreamt of starting a lab where wild dolphins could come and go freely while researchers worked at communicating with them, but lacked the funds to try it. So he returned to research on human consciousness, taking LSD in isolation tanks, and experimenting with the effects of a little-known drug called ketamine. In the process, he moved to California, where, it seems, cetacean consciousness pursued him.

The actor Burgess Meredith wanted to make a dolphin film and invited Lilly to stay at his beach house while he was looking for property in Malibu. The next morning, according to Lilly's biographer, Meredith told him that he'd dreamt that a dolphin came ashore under his house, and that his wife and some neighborhood children cooperated to return it to the water. "It was a strange dream for Burgess; he'd never actually seen anything like that. Two hours later a small commotion erupted beneath the Meredith home, built on poles over the beach. . . . There in the surf was a dolphin, swimming ashore. It beached itself, lying on the sand." Cooperating with neighborhood children, as in the dream, Meredith's wife returned it to the water.

"Such scenes with dolphins and whales occurred frequently around John," his biographer noted. "Whenever he was at the beach, the creatures were certain to make an appearance in the water. If he went to a marine park, the orca acknowledged his presence by refusing to perform. He seemed to have some mysterious connection with cetaceans."

Lilly decided to try again at communicating with dolphins. With his third wife, Toni Ficarotta, he created a computer system that he thought might translate their language. Working with free ones remained impossible, so they captured two young bottlenoses, named them Joe and Rosie, and kept them in a lab in the back lot of Marine World/Africa USA in Redwood City near San Francisco from 1979 to 1985. The project further convinced Lilly of cetacean powers: "He was coming to the conclusion that through their vocalizations dolphins create a three-dimensional world of sound, a virtual reality of sonic images resembling a kind of 'cetacean television' constantly broadcast at one another. . . . The situation is somewhat analogous to what might happen if humans could communicate by creating images directly in one another's minds or, perhaps, to the world of images humans generate through electronic communication."

Lilly hoped he "might be able to disconnect his internal reality from the external reality through isolation, and connect it with the virtual reality of the dolphins," and he had experiences that he thought approached that goal. But the project did not progress that much farther toward communicating with dolphins in "external reality" before it ended because of funding and computer problems: "By the time it was wrapped up, Joe and Rosie, approaching sexual maturity, knew high pitched, computerized versions of fifty human words, and, more significantly, the syntax to go with them. . . . [T]hey caught on to the idea that the precise *sequence* of words conveys additional meaning in the human world, and they used the sounds and sequences deftly to respond to humans and control events in their environment."

After their release, Joe and Rosie were seen swimming with a calf, but they were not heard from again. By then, the work's slow progress had exhausted Lilly's patience with earthly external reality. In a 1988 speech, he said that he had been trying to escape his "body and limited reality" since his isolation tank and LSD experiments decades earlier. In 1990, he urged society "to learn to communicate with large-brained extrahumans—dolphins, whales, and elephants—so that we will be prepared to communicate intelligently with extraterrestrials."

Lilly's cosmic aspirations seem to have limited his interest in nonhuman life. He ignored cetaceans with a smaller brain-to-body ratio than humans because he thought they would lack the ability to "communicate complex thoughts." He tentatively classed gray whales and harbor porpoises, not to mention sea lions, seals, and otters, as "animals," and sperm whales, orcas, and bottlenose dolphins as "humans." Such a classification was speculative, however. Baleen whales may be less capable than dolphins or sperm whales of Lilly's "complex communication" because their brains are not as big in ratio to body size and because they lack equipment for echolocation. But there is no proof. (Annalisa Berta has noted that, judging from mitochondrial DNA, some researchers think sperm whales may be more closely related to baleen whales than to other toothed whales.)

Many experienced observers have not made Lilly's distinctions. Scammon considered gray whales particularly smart because of their behavior in the Baja lagoons. "When the parent animal is attacked, they show a power of resistance and tenacity of life that distinguish them from all other Cetaceans. Many an expert whaleman has suffered in his encounter with them, and many a one has paid the penalty with his life." Tenacious resistance seems predictable in the circumstances, but Scammon thought it remarkable compared to most other whales—including sperm whales, the possessors of Lilly's "huge computer."

Scammon made sperm whales sound like Steller's dim-witted sea cows when under attack: "[T]hey all 'bring to' and remain, usually for some time, with their dying companion," he wrote of females, "by this means a number of whales are often captured from the same school." He thought males less self-sacrificing, but still not very smart about resisting whalers' attacks: "[I]f one of their band is harpooned, its cowardly associates make off in great trepidation. When individually attacked, however, it makes a desperate struggle for life, and often escapes after a hard contest. Nevertheless, it is not an unusual occurrence for the oldest males to be taken with but little effort on the part of the whaler."

A recent development may support Scammon's estimation of gray whale sagacity. They retained a reputation for ferocity even after they were protected, allegedly attacking expeditions that came simply to study or photograph them. Then, in the 1970s, when whale watching had replaced whaling on most of the Pacific coast, they suddenly became friendly, mothers bringing their calves to tourist boats to be petted. Apparently, as with the story Lilly heard of orcas and harpoon boats, the news that tourist boats are harmless had spread through the population.

Another manifestation of baleen whale communication emerged in 1967, when two biologists, Roger Payne and Scott McVay, discovered that haunting underwater noises made by male humpbacks have intricate patterns. Listening through hydrophones, they found that the whales repeat long sequences of sounds as birds repeat their songs. "These songs are much longer than birdsongs and last up to thirty minutes," Payne wrote. "They are divided into repeating phrases called themes. When the phrase is heard to change (usually after a few minutes), it heralds the start of a new theme. Songs contain from two to nine themes and are strung together without pauses so that a long singing session is an exuberant, uninterrupted river of sound that can flow on for twenty-four hours or longer."

The biologists found that the whales sing only in breeding waters, that they sing different songs in different oceans, and that the songs change gradually as individuals develop and introduce new variations: "Humpback whales change their songs continually so that after about five years they are singing an entirely new song, and apparently do not ever return to the original." According to Payne, humpback songs are like human ones in that they contain rhythms and rhymes, which suggests a high level of communication.

Even pinnipeds show a potential for "complex thoughts." In 2005, the media made much of a young captive orca who had learned to use regurgitated fish on the surface of his tank as bait to catch gulls, then passed the trick on to others. But Scammon observed similar behavior in wild California sea lions:

> When in pursuit, the animal dives deeply underwater and swims some distance from where it disappeared; then rising cautiously, it exposes the tip of its nose above the surface, at the same time giving it a rotary motion, like that of a water bug at play. The unwary bird on the wing, seeing the object near-by, alights to catch it, while the Sea Lion at the same moment, settles beneath the waves, and at one bound, with extended jaws, seizes its swimming prey, and instantly devours it.

I haven't seen sea lions do this, but the alacrity with which murres flush away from passing ones is suggestive. (Sea otters also occasionally ambush birds, although I haven't seen an account of them using bait.)

Pinnipeds showed unexpected communication potential when a harbor seal in the New England Aquarium began accosting visitors with remarks like "Hello there," "Hey," "How are you?" "Get out of there," and "Come out of here." The seal, named Hoover, also made a laughing sound, interpreted as a repetition of his "Hey" sound: "heyheyheyheyhey." The words were "strung together in a nonsensical way" and were often the same ones used by caretakers. Still, considering Lilly's laborious dolphin training, Hoover's spontaneous loquacity is affecting. I heard him on the radio. He sounded like a bored, slightly drunk man with a Boston accent shouting: "Hoovah! Hoovah! Come ovah heah!"

Found as an orphaned pup in Maine in 1971, Hoover was a pet for three months before he arrived at the aquarium. There, it was reported,

[his] development was unremarkable until the summer of 1973, when he began to display "spasms," "shaking," and "raspy breathing." This continued in 1974, along with "chest-slapping," "strange cries," and "raspy growling sounds." Hoover was ill during some of this period, exhibiting congestion, vomiting, dehydration, and weight loss, and much of his strange behavior was thought to be associated with his illness. However, several observers noted in the files that he seemed to stage "attacks" to get human attention. In 1975, Hoover was extremely vocal, particularly in May, June, and July, and in 1976 he exhibited sexual behavior for the first time. In August, 1976, an observer noted that Hoover made sounds "as if talking." He continued to make "new noises" and developed a "blood-curdling scream" in 1977.

Hoover was first heard to mimic a human word in November 1978, when he was seven years old. One observer wrote in the files: "He says 'Hoover' in plain English. I have witnesses." A volunteer working with the seals subsequently reinforced the vocalization at times by blowing a whistle and rewarding Hoover with fish when she heard him "say his name." His repertoire of speechlike sounds increased rapidly but these sounds were not developed by shaping on the trainers' part.

Biologists guessed that Hoover was mimicking human speech instinctively in the absence of sounds male harbor seals normally learn from other males. But this didn't explain how he learned to do it, assuming an odd posture with his head thrown back, nose pointing up, and neck retracted. It didn't explain why he took seven years to learn, although the onset of sexual maturity evidently played a part. No researcher with the time and independence of John Lilly studied Hoover's behavior, and it remains mysterious.

Interpretations of apparent cetacean communication like gray whales' switch from ferocity to friendliness also amount to little more than guesses. The switch might seem self-evident behavior now that humans have stopped killing grays and started admiring them. It might seem evidence of short memory. But that interpretation is probably superficial given the complexity of other whale behavior.

Humpback whale singing certainly is not obvious and has been interpreted as religious chanting, storytelling, epic poetry, symphonic music, auditory mapmaking, and, as with birds, a way for males to attract females and compete with other males on breeding grounds. Roger Payne inclined toward the latter interpretation. "I get the impression," he wrote, "that the main difference between the songs of humpbacks and songbirds is their pace and pitch—a consequence of each animal's size." But he acknowledged that this was speculation:

> Since 1967 I have listened on and off to the songs of humpback whales—researching them, analyzing them, ransacking through them, beseeching them, always yearning to know what they mean. But they guard their secrets as effortlessly, as enigmatically, as they always have. . . . I frequently get asked to comment on the mental abilities and communication skills of large whales and have to invent new ways to explain that we simply have no evidence one way or the other with which to answer these questions.

As Lilly perceived in the 1950s, the best way to understand the communication of nonhuman organisms would be to learn their "languages." But we are no closer to that than his lab was. One reason he didn't get very far, at least in "external reality," might be that the task implies not only a knowledge of the creatures' brains and sensory organs, but of their lives. For all his brilliance, Lilly had a certain blindness to the ecosystems behind the brains he wanted to contact. His books never described seashores or seascapes and showed little interest in the evolutionary past, or even in living dolphins' natural history. In this, he resembled the military-industrial complex whose preemption of cetacean communication he feared. Navy dolphins trained to "watchdog" ships off Iraq are part of his legacy, along with academic studies of cetacean psychology.

That perhaps was one reason why marine biologists mistrusted Lilly's influence. As Roger Payne wrote in his 402-page *Among Whales,* which made no reference to Lilly or his dolphin research: "Whales have had their most advanced brains for almost thirty million years. Our species has existed for about one-thousandth of that time. We have a lot to learn from whales." Payne added that what impressed him most about whales was the slowness of their lives. Lilly's attempts to communicate with them at high-tech speed seem superficial in contrast.

One might say that the most impressive thing about life in general is its slowness compared to human culture. The past 500 million years on North America's west coast, a one-eighth slice of the 4-billion-year pie, are hard to imagine. Even the past 65 million years during which our class Mammalia has dominated the coast are hard to imagine compared to the 10,000 or 20,000 during which humans have lived here. They are, in fact, unimaginable to most Americans, who cling to a biblical 10,000 years.

Still, the evidence is in museums. It would be easy now for Rudolph Zallinger to do a Cenozoic counterpart of his marine reptile painting, using Emlong's late Oligocene to mid-Miocene fossils. It would be a darker blue-green than the reptile painting because of the sea grass and kelp in the water. The shore, lined with pines and oaks instead of cycads, would be lower and shallower to show *Behemotops* grazing, *Kolponomos* clamming, *Aetiocetus* stirring the bottom for crabs, and *Enaliarctos* climbing out on a rock to eat a lamprey. But the offshore water and sky would be just as expansive and full of giants: pterosaur-like *Osteodontornis* soaring overhead; *Hesperornis*-like *Tonsala* paddling and diving; white sharks and a mosasaur-like *Macrodelphinus* cruising after unwary members of a seal-like *Desmatophoca* herd swimming past.

As with the marine reptile painting, the panorama would give just a glimpse of the coastal story rather than the *Age of Mammals'* more extensive land beast narrative. But it would still imply evolution, from coast-hugging *Enaliarctos* to pelagic *Desmatophoca,* from toothed *Aetiocetus* to baleen-bearing cetotheres blowing on the horizon.

Such visions stir the imagination. They also resist it. Geological time weighs heavily against dreams of quickly imposing human culture on evolution in more than rudimentary, blundering ways like cloning or recombinant DNA. As Payne pointed out, cetacean mentality, whatever it is, stems from an unimaginably longer period of exceptional brain growth than ours. When Jacques Yves Cousteau suggested that orcas might be given hands through genetic engineering and thus transformed into undersea technologists, he echoed Emlong's pseudoscientific fantasies and Lilly's impatient speculations. Tool-using sea mammals already exist, anyway. I once read a science fiction story wherein humans, having made land uninhabitable, choose sea otters as successors and, to help them evolve, erect beach monoliths bearing all knowledge condensed to equations. The narrator watches an otter come ashore and scratch at one. Examining it afterward, he finds some equations corrected and new ones added.

Learning to communicate with orcas well enough even to propose giving them hands might take countless years of hard work and big budgets, if it is possible. Perhaps if enough people learned to imagine 65 million years, it could be tried. Life may or may not exist in outer space: it certainly exists in sea creatures' brains.

I remember my impatience when I was in southeast Alaska in 1972 and didn't see any more marine fauna than John Muir had a century before. There was so much water in all those fjords, and so little life that I could perceive. I finally saw some humpback whales when I spent several days in an abandoned Forest Service cabin on Taku Inlet south of Juneau, but they were miles out toward Admiralty Island and looked minnow-sized. I could just see them, spouting, breaching, and waving their flukes in the air, which was very still and clear. They were so distant that long seconds would pass between the sight and the sound.

It was June, and dusk lasted until nearly midnight, with gulls and bald eagles circling in a red sky as the whales continued their ponderous but apparently minuscule activities. Not being used to the long twilights, I felt transported out of normal time. When it finally got dark, for some acoustic reason or perhaps just because I could no longer see them, the whales began to sound startlingly close, as though they were spouting and breaching over the tide pools, wildflower patches, and Sitka spruces around me. I felt their giant presence over my head. It was like something in "Pileated Woodpecker Crosses the Ocean."

At the time, I had no idea what the humpbacks were doing. Now I think they may have been "bubble netting" to concentrate fish schools so they could engorge them in their pleated throats. I've since seen underwater films of them doing this and other things, and I've heard recordings of their songs. I still feel that what I really know about humpbacks is what I saw and heard that evening at Taku Inlet.

The Old Man of the Sea

When carrion birds gather on a beach near Point Resistance, I feel anticipation mixed with trepidation. I want to see what has come ashore, but I know it will be gruesome, like a pile of decapitated gulls I found after one storm, or sad, like the harbor porpoise calf with its empty eye sockets. And sometimes it is not even informative. One object that attracted vultures after an autumn storm was a fibrous brown mat from which protruded odd bones and organs—ribs, a kidney, a trachea. The trachea must have been part of a neck, but no skull was attached, although there may have been whiskers. Excrescences at the mat's edges might have been flippers or fins, but they had no bone or ray structure.

I decided it probably had been a half-grown California sea lion. A disease, leptospirosis, had been killing them, and a recognizable carcass came ashore soon after. But that was a guess. For all I knew, the mystery mat might have been a Steller's sea ape.

Less notorious than those having to do with Bigfoot, west coast sea monster legends are similarly persistent. Richard Ellis, who puzzled over the sea ape in a 1994 book, puzzled again in two more. In 2001, he repeated his idea that the creature might have been the extinct Atlantic sea mink's Pacific counterpart:

> [A]lthough there is no reason to assume that a comparable species existed on the West Coast of North America or the East Coast of Asia, there is an

Alaskan subspecies of the common mink (*M. vison injens*) and several species of martens, sables, and other mustelids live in Siberia. In other words, if Steller had spotted the "sea monkey" on the East Coast of North America, there would be no problem with its identification.

Ellis seemed more tentative than before, however, concluding: "There are no fossils that might confirm the existence of such a creature." In 2003, he dropped the sea mink idea and allowed that "what Steller saw that evening" will probably remain unknown. But, he continued, "a clue has surfaced that would seem to tilt the interpretation toward an unknown animal." He quoted from a 1715 Japanese book with a picture of a sea ape-like creature: "The color of its fur resembles that of a fox, and the shape of its tail is similar to a fish's. Its legs resemble a dog's, but it has no forelegs. . . . The whole body is that of a fish, and it has a tail, but it is half fish and half beast. Its head resembles a cat's, and its muzzle is sharp. It has eyes and a nose, but no external ears, only small holes."

Ellis acknowledged that two biologists, Victor Scheffer and Charles Greer, had interpreted this as a "perfectly good description of a northern fur seal, which, they say, was known to Asiatic peoples . . . long before Europeans came to the North Pacific." But he thought it sounded more like a sea ape, except for its lack of external ears. "Perhaps," he mused, "there were marine apes with ear flaps, a la Steller, or others with just ear holes."

Like Bigfoot, a nocturnal hominid at home in alpine snow, the sea ape would be a biological anomaly. In Steller's description, it seems mammalian, but no known mammal lacks forelimbs, although some have superficially fishlike tails. It also seems intelligent, which might tempt partisans of Hardy's sea ape to see it as that hypothetical primate's distant marine descendant. But 5 million years is a very short time to produce a creature so different from an African anthropoid. It took longer for sirenians to evolve from subtropical *Dusisiren* to subarctic *Hydrodamalis,* and the changes they underwent, while great, were less than evolving from a four-limbed ape to a limbless, fishtailed one. Anyway, the doglike head, ears, and whiskers Steller described make it seem an unlikely ape descendant. It doesn't seem cetacean either, which leaves it out of John Lilly's other possible "human" category. It sounds more like a pinniped or sirenian, but not very much.

Even among legends, the sea ape is anomalous. It is not sexily humanoid, like a mermaid. It is not impressively gigantic, like Bigfoot or a sea serpent. Its significance remains obscure. In that obscurity, as well as its mythic quality, it reflects my dream of sea creatures coming ashore. I don't know why that dream was so uniquely vivid and memorable. Such qualities imply a meaning, but I'm not sure what it is. Dream meaning is elusive.

A recent common scientific view maintains that dreams are random neural activity, without intrinsic meaning, so that interpreting them is like drawing horoscopes from the stars. Most people have considered dreams meaningful, however. Prehistoric cultures regarded them as portals to a spirit world. Dreamlike images in Native American stories reflect this—Tom Smith's souls drawn west through the surf, Tony Bell's multicolored banks of sea creatures, John Sky's iridescent blue-green beings. Historic cultures continued to see dreams as supernatural portals. Charles Sternberg, Edward Cope's assistant, said that during his Kansas youth he dreamt of fossils on the prairie and later found them in the place he'd dreamt. Modern cultures began to see scientific meanings in dreams. Early psychologists like Freud thought they might be one way that the unconscious—a mind underlying rational waking consciousness and "thinking" in more instinctual ways—interprets personal experience.

I don't know about spirit worlds, but my dream clearly arose from personal experience. Florida dolphins and pelicans enthralled me in childhood, as did Rudolph Zallinger's marine reptile painting. So did Rudyard Kipling's *Jungle Book* story "The White Seal," about a fur seal that escapes from the slaughter on the Pribilofs and wanders in search of refuge, than encounters a herd of Steller's sea cows and follows them to a secret lagoon, accessible by an underwater cave, where sea creatures can live safely. Later, I found watching harbor seals on the Maine coast and reading Robert Merle's thriller *The Day of the Dolphin,* in which communication with dolphins saves humanity from nuclear annihilation, just as fascinating.

Still, that doesn't explain why it was my most memorable dream. Experiences of greater personal significance haven't produced such a memorable dream. And I kept having dreams of sea creatures coming ashore. For a while, they grew increasingly spectacular and colorful, with creatures shooting into the air and whizzing past me as I stood on high places. Sometimes I flew over cliffs and shores myself. The dreams settled down later but continued to play on my experience.

While researching this book, I dreamt I was on an expedition to a place where a whale-sized, tadpolelike creature had come ashore to a murky tide pool, browsing algae from the rocks. An expedition member swam into the pool, where she saw not the algae browser, but a small blue and white orca, which she said was extinct. She then proposed another expedition to find the tadpole-whale but said (random neurons at work?) I would have to change my brand of shampoo if I wanted to go.

Some early psychologists like Jung thought there is more to dreams than personal experience, that they express things inherent to mind, archetypes. One implication of this idea is that archetypes might not be confined to human brains but might connect to others. Although indifferent to whales on their Baja expedition, John Steinbeck and Ed Ricketts thought along these lines. The ocean, they wrote, "deep and black in the depths, is like the low dark levels of our minds in which the dream symbols incubate and sometimes rise up to sight." They speculated that the mind recapitulates prehuman evolution in the womb, leaving the individual with an unconscious memory of sea life, and that marine phenomena like the tides thus influence the adult: "If the gills are a component of the developing human, it is not unreasonable to suppose parallel or concurrent mind or psyche development. If there be a life memory strong enough to leave its symbol in vestigial gills, the predominantly aquatic symbols in the individual unconscious might well be indications of a group psyche-memory which is the foundation of the whole unconscious. And what things must be there." One of the things Steinbeck and Ricketts thought must be there was a creature called the Old Man of the Sea by Monterey fishermen, who described him as looking "somewhat like a tremendous diver, with large eyes and fur hanging shaggily." They mused:

> When the fishermen find the Old Man rising in the pathways of their
> boats, they may be experiencing a reality of past and present. This may
> not be a hallucination; in fact, it is little likely that it is. The interrelations
> are too delicate and too complicated. Tidal effects are mysterious and
> dark in the soul, and it may well be noted that even today the effect
> of the tides is more valid and strong and widespread than is generally
> supposed. . . . The imprint lies heavily in our dreams and on the delicate
> threads of our nerves, and if this seems to come a long way from sea-
> serpents and the Old Man of the Sea, actually it has not come far at all.
> The harvest of symbols in our minds seems to have been printed in the
> soft rich soil of our prehumanity. Symbol, the serpent, the sea, and the
> moon might well be only the signal light that the psycho-physiological
> time-warp exists.

Such thinking is poetic, of course. It can mislead someone like Douglas Emlong, who applied it literally to a fantasy of cybernetic neo-Lamarckism:

The Unconscious Mind may have many of its roots in the bioplasma, and the evolution of life may be motivated by incentives for change, somehow recorded in the mind matrix, and the DNA and RNA genetic codes. Life units resemble machines, and when the equilibrium of a species is impaired repeatedly under survival conditions, by a weak component, this may be recorded by the Unconscious Mind, which may trigger a trial mutation by a computer-like process, such as will take place in self-duplicating machines, which will improve themselves in the near future. This factor may accompany Darwinian evolution.

Emlong's notion of the unconscious as an entity that can trigger genetic mutations is unfalsifiable pseudoscience, especially given our sketchy understanding of the brain-mind relationship. But the notion of dreams as random neuronal activity may be unfalsifiable, too. Philip K. Dick, a writer who fictionally played with ideas that Emlong took literally, made a perceptive comment: "The unconscious relates to the entire world, the whole panoply of the universe. It learns from birds, . . . it learns from everything at once." Perhaps the unconscious is the whole of interactions between a nervous system and its environment, the foundation on which consciousness develops. In this sense, it might be, as Emlong naively proposed, an evolutionary basis of learning, and thus a source of adaptation.

Scientists disagree as to whether nonhuman animals have conscious minds. But if there is such a thing as the unconscious mind, nonhuman animals have it. They also learn "from everything at once," and they manifest their learning in ways we can understand. Gray whale calves spyhop like human infants peeking out of strollers as their mothers lead them for the first time past the murre-covered rock at Point Resistance. Elephant seal pups play in Drake's Bay as I did in the Florida surf as a child, letting the waves push them up and down the shallows. Young bulls linger at the rookery after the beach masters leave, roaring feebly, as though pretending to be eighteen feet long.

We share another aspect of the unconscious mind with nonhuman creatures. Other mammals, at least, dream. It seems likely that their dreams play a part in their emotional lives similar to the part they play in ours, and perhaps they have an imaginative side as well. Maybe they also dream of flying, which might be why sea mammals like to jump. Maybe seals and porpoises dream about land creatures coming to sea in these days of surfers and scuba divers.

Of course, we have no idea what they dream. Most west coast sea creatures might as well be in the Pliocene for all the apparent relation they have to humans. Watching them feed on a herring run off the ocean beach at Point Reyes, I get the same ghostly feeling as with the humpback whales at Taku Inlet. There's great excitement out there. Sea lions line up with their flippers in the air; harbor porpoises dive busily in a tight circle; cormorants paddle around in another; pelicans hover and dive. But from the beach it's just a tiny frieze on the horizon.

Still, creatures do take notice. Harbor seals keep a careful eye on the land. Once when I was on a cliff above a beach full of the usual obliviously sleeping elephant seals, I noticed two smaller pinnipeds and turned my binoculars on them. Instantly, a harbor seal mother saw I was watching and edged her pup toward the water. It is hard to walk along any beach at Point Reyes without a harbor seal watching at some point, and at twilight surprisingly large groups of them may surface very close to the beach and watch, craning their necks curiously.

Such notice can take surprising forms. A 2005 *San Francisco Chronicle* article on Golden Gate Bridge suicides mentioned that a man who had jumped felt a creature "circling him, nudging him, preventing him from sinking back into the horrifying darkness" long enough for the Coast Guard to pick him up. The jumper thought it was a shark, but a spectator photographed a sea lion near him. Another *Chronicle* article that year about the rescue of a humpback whale tangled in crab pots did not mention that a pinniped had been leaping around the whale, but TV news video showed what a diver said was a harbor seal doing so. When freed, the whale "swam to each diver, nuzzled him, and then swam to the next one." (The diver didn't say if it nuzzled the seal, too.)

At Point Resistance, vultures and gulls glance at me as they pass overhead, and sometimes circle back curiously. The most persistent noticer there, unsurprisingly, is the star of Haida creation stories. Ravens may seem preoccupied, as when they amuse themselves by hovering in cliff edge updrafts, rising or sinking as if on invisible elevators, but they are watching everything. Sometimes when I leave, they fly down to where I was sitting to see what I was up to.

Nonhuman life is not opaque and static, as it may seem. It is aware and always changing, albeit often imperceptibly slowly or quickly. (A week after the record heat wave of July 2006, boaters saw a large manatee in New York's Hudson River.) Such qualities suggest a meaning for my dream. Its main impression was of reassurance, as though the dream manifested a core of unconscious vitality that emerges to renew the apparent stony vacancy of everyday existence, rather as the planetary core's heat tectonically renews the stony crust. It seemed a token of continuance, of Pileated Woodpecker's "going yet," which seems improbable in an incomprehensibly dangerous universe but somehow occurs.

A few years ago I dreamt of standing on a skyscraper, looking north at a body of water that extended to the horizon. It was spreading south, engulfing the city, and people were embarking from the drowning parts in small boats. Some began diving, and I could see them descending deep, although they were far away. When they surfaced again, they had become seals, like harbor seals, but different—perhaps like the seals Scammon saw at Point Reyes in 1855. I watched them swim ashore, and soon many were basking on a forested point.

It was a less reassuring dream than the original one, but it still manifested a thread of continuance. Whatever happens in the next few thousand years, North America will continue to meet the Pacific. That meeting may be far east of its present location if climatic warming raises sea level. It may lack its present life. But there will be tetrapods, and the interactions of land and water will continue.

After I visited Emlong's fossils at the Smithsonian, I went to Calvert Cliffs, the famous Miocene deposit on Chesapeake Bay. I didn't see living sea mammals, just scraps of seal or whale bone. But I saw a possibility of continuance. The trail to the bay ran down a creek that beavers had dammed surprisingly close to the shore, making a big pond covered with yellow-flowered spatterdock and fallen trees on which countless shiny black turtles sunbathed. There must be good reasons why rodents, the most prolific and adaptable mammals, have never gone to sea. But that doesn't mean they never will. The place, full of reflecting sunlight as the wind shook the spatterdock leaves, seemed to evoke emergent worlds.

The cliffs at Point Resistance also overlook a tiny creek mouth where the wind shakes the leaves, and although there are no beavers, there is still a sense of possibility. The place's name might refer to life's emergent qualities, its resistance—as spirit or mind or whatever it is that fosters continuance—to the physical world's entropic randomness. When I asked the National Seashore archivist what the name derives from, she said nobody in the park seemed to know. Like a dream, it must have a source, but it is obscure, so it might as well stand for continuance.

The west coast's sea creatures embody continuance, not just in their past survivals but in their present lives. And many, of course, do come ashore to bring life to lonely, stony places—like the nesting murres at Point Resistance and the breeding harbor seals at their spectacular beach south of it. Watching elephant seals do so at their unspectacular beaches northwest of it is like a time-lapse version of my dream.

For months only boulders mark the sand. Then, one November day, an elongated gray-brown shape appears that might be a boulder if a giant snail track did not extend from the surf, and if it did not sometimes raise one end and open surprisingly large, lustrous black eyes. The next day, the bull is gone, but another appears the day after, yawning and rattling his proboscis with a noise like blocked drains. A week later there are two or three bulls, and some tan females. A few weeks later, there are dozens of females, and the wailing black pups appear. Harbor seals play offshore, sometimes jumping, apparently infected by the excitement. Once a slender, golden sea lion female perched right in the rookery's center, as though basking in the procreative ambience.

One early January, I saw a female with a newborn pup at a small beach on the ocean side. Most of the seals pup on the calmer Drake's Bay side, and doing so on a strip of sand between Pacific surf and a cliff seemed precarious. The next week, a storm raised the highest tides in a long time, and I thought the seal and her pup would be swept away, fatally for the pup, since newborn elephant seals don't swim. I expected an empty beach when I returned later in January. Instead I saw ten pups, one large enough to be the newborn of two weeks before. Another small ocean beach, pupless on my earlier visit, had four, one nursing at a mother with a healed shark bite beside her teat.

A month later, the second ocean beach was empty, and there were fewer seals on the first. So perhaps storms had swept some pups away. But the sea sweeps to death and life impartially. After finding the gruesome mat that might have been a half-grown sea lion at Limantour beach, I walked to Point Resistance. As I got there, a half-grown sea lion was swimming north between the murre rock and the cliff. The day was windy, and choppy waves knocked the little creature around. But it kept on, probably headed for the basking rocks below Point Reyes, and was soon out of sight.

NOTES

PROLOGUE. STELLER'S SEA APE

Page

xix   "In 1999, passengers on a Farallon tour boat": Klimley, *The Secret Life of Sharks,* p. 29.

"In 1880, a steam launch": Voy, "Blue Whale Hunt in Drake's Bay," p. 295.

xx   "The animal was about two ells": Steller, *Journal of a Voyage with Bering,* p. 82.

xxi   "In a treatise on marine mammals": ibid., p. 24.

xxii   "Walking on the shore": ibid., p. 128.

"The largest of these animals": ibid., p. 160.

"These animals are found at all times": ibid., p. 162.

xxiii   "When I had observed it ": ibid., p. 83.

"Most of the peculiarities of behavior": Stejneger, *George Wilhelm Steller,* p. 280.

"Victor B. Scheffer, a marine mammal biologist": Scheffer, *The Year of the Seal,* p. 6.

xxiv   "On August 4, when sailing to the south": Steller, *Journal of a Voyage with Bering,* p. 4.

"Corey Ford, author of a 1966 book": Ford, *Where the Sea Breaks Its Back,* p. 91.

xxiv   "But then, as the Petrograd copy's editors plaintively ask": Steller, *Journal of a Voyage with Bering,* p. 199.

       "Ford thought that 'the simplest explanation is that the "sea monkey" actually existed'": Ford, *Where the Sea Breaks Its Back,* p. 91.

       "In a 1994 book": Ellis, *Monsters of the Sea,* footnote to p. 98.

       "Thomas Pennant, an eighteenth-century naturalist": Matthiessen, *Wildlife in America,* p. 85.

       "He quoted two Canadian zoologists": Ellis, *Monsters of the Sea,* p. 69.

       "Querying marinas, lighthouses": ibid., p. 70.

       "Ellis included a fuzzy 1937 photo": ibid., p. 71.

xxv   "Heuvelmans, in his voluminous review": LeBlond and Bousfield, *Cadborosaurus,* p. 74.

       "LeBlond's book reproduced clearer photos": ibid., p. 51.

       "According to an eyewitness": ibid., p. 55.

       "The station's annual report": ibid.

       "People have reported": Reinstedt, *Mysterious Sea Monsters of California,* pp. 22, 46.

       "Peter Simon Pallas, a Russian naturalist": Steller, *Journal of a Voyage with Bering,* p. 29.

xxvi   "John Steinbeck and his marine biologist friend": Steinbeck and Ricketts, *The Log from the Sea of Cortez,* pp. 30, 31.

       "I frequently observed": Steller, *Journal of a Voyage with Bering,* p. 81.

xxvii   "In 1931, a Miwok man": Kelly, "Coast Miwok," in Heizer, *Handbook of North American Indians,* vol. 8, p. 423.

xxix   "[A]lthough the *San Francisco Evening Bulletin* reported": "Limulus in the Pacific," *American Naturalist* 20 (1886): 654.

CHAPTER I. REEFS IN THE DESERT

1   "The scene was dominated": Schopf, *Cradle of Life,* p. 90.

2   "In western North America, the 850-million-year-old Kwagunt Formation": ibid., pl. 1.

       "Charles D. Walcott, a paleontologist with the U.S. Geological Survey": ibid., p. 27.

       "In 1996, Mark McMenamin": McMenamin, "Ediacaran Biota from Sonora, Mexico," p. 4990.

4 "Stephen Jay Gould, the late twentieth century's paleontological superstar": Gould, *Wonderful Life,* p. 23.

"Gould accordingly claimed it as 'another Burgess oddball'": ibid., p. 193.

"In 1990, however, a researcher": Conway Morris, "Showdown on the Burgess Shale," p. 49.

5 "From the dawn of the Cambrian": Barnett, *The World We Live In,* p. 94.

6 "Indeed, one theory of the origin of hard parts": Maisey, *Discovering Fossil Fishes,* p. 41.

7 "[I]n the 1980s, eastern North American fossil soils": Gordon and Olson, *Invasions of the Land,* p. 110.

"For perhaps a billion years": Barnett, *The World We Live In,* p. 96.

8 "A botanist, G. Ledyard Stebbins, thought": Stebbins, *Darwin to DNA, Molecules to Humanity,* p. 225.

"Derek Briggs, who studied the Ordovician animals": Wade, *The Science Times Book of Fossils and Evolution,* p. 116.

CHAPTER 2. AMPHIBIOUS AMBIGUITIES

10 "According to *The World We Live In*": Barnett, p. 94.

12 "To this day I do not know": Klimley, *The Secret Life of Sharks,* p. 49.

13 "These fins, and fossil evidence": Maisey, *Discovering Fossil Fishes,* p. 79.

16 "In 2006, they announced": Daeschler et al., "A Devonian Tetrapod-Like Fish," p. 757.

18 "The Bear Gulch Limestone is 'unique'"; Lund and Poplin, "Fish Diversity of the Bear Gulch Limestone," p. 286.

"The reptiles, and remains of other organisms": Gordon and Olson, *Invasions of the Land,* p. 155.

19 "Fossil sediments suggest that atmospheric oxygen levels dropped": *San Francisco Chronicle,* April 15, 2005.

CHAPTER 3. BIRD TEETH AND REPTILE NECKS

20 "Edward D. Cope, the pioneer paleontologist": Ballou, "The Serpentlike Sea Saurians," p. 223.

21 "We know that marine crocodiles occurred": Hilton, *Dinosaurs and Other Mesozoic Reptiles of California,* p. 66.

22 "Cuvier likened Lamarck's ideas to 'enchanted palaces'": Wallace, *Beasts of Eden,* p. 12.

23 "In an 1841 paper": McGowan, *The Dragon Seekers,* p. 180.

25 "On an 1872 expedition": Marsh, "Odontornithes, or Birds With Teeth," p. 626.

"That *Hesperornis* was carnivorous": ibid., p. 629.

26 "The first species of birds": ibid., pp. 625–26.

"The fortunate discovery of these interesting fossils": Schuchert and LeVene, *O. C. Marsh,* p. 426.

"Nothing so startling has been brought to light": ibid., p. 427.

"Such a discovery": "Another Bird with Teeth," *New York Herald,* September 21, 1876.

27 "The classes of Birds and Reptiles": Marsh, "Introduction and Succession of Vertebrate Life in America," p. 352.

"On receiving one, Huxley wrote": Schuchert and LeVene, *O. C. Marsh,* p. 232.

"The vertebrae of the sea snakes": Zeigler, "The Rocky Mountains," *New York Herald,* December 24, 1870.

28 "Seeing Cope's reconstruction": *New York Herald,* January 19, 1890.

"Marsh has been doing a great deal": Osborn, *Cope,* p. 160.

"If Marsh had found a pterosaur": ibid., p. 166.

"In 1867, he had written": Cope, "Fossil Reptiles of New Jersey," p. 25.

29 "It is known": Cope, "The Necks of the Sauropterygia," p. 132.

30 "Although able to 'tell chalk from cheese'": Webb, *Buffalo Land,* p. 38.

"One of his Kansas collecting assistants": Sternberg, *Life of a Fossil Hunter,* p. 69.

"Every animal of which we had found trace": ibid., p. 75.

"These strange creatures flapped": Lanham, *The Bonehunters,* p. 98.

31 "A *New York Times* editorial": *New York Times,* September 7, 1877.

"The *Tribune* described it": *New York Tribune,* October 10, 1877.

"The *New York Herald,* mocker of Marsh's sea serpent bones": *New York Herald,* October 9, 1877.

"In it, Cope accused Marsh": *New York Herald,* January 12, 1890.

"Williston had begun as a Marsh admirer": Schuchert and LeVene, *O. C. Marsh,* p. 186.

"The larger part of the papers": *New York Herald,* January 12, 1890.

"Cope insisted, however": *New York Herald,* January 20, 1890.

32 "The Philadelphia Academy's Joseph Leidy": *Philadelphia Inquirer,* January 13, 1890.

"Cope, perhaps, defined the greatest number of species": Ballou, "The Serpentlike Sea Saurians," p. 213.

"According to Ballou, Marsh called on the editor of *Popular Science Monthly*": Wallace, *The Bonehunters' Revenge,* p. 278.

"After receiving a copy of the deluxe toothed bird monograph": Schuchert and LeVene, *O. C. Marsh,* p. 247.

"Even the die-hard anti-Darwinian Richard Owen": McCarren, *The Scientific Contributions of Othniel Charles Marsh,* p. 41.

CHAPTER 4. TAIL TALES

33 "In 1893, James P. Smith": Hilton, *Dinosaurs and Other Mesozoic Reptiles of California,* p. 129.

36 "Something that Merriam and Alexander did find": ibid., p. 80.

37 "Fragmentary pterosaur fossils": ibid., p. 71

38 "If so, it may have fed heron-fashion": Kevin Padian, phone interview, March 14, 2005.

"Kevin Padian, a paleontologist at UC Berkeley": ibid.

"Fossil nothosaur embryos not enclosed in eggs": Renesto et al., "Nothosaurid Embryos from the Middle Triassic of Northern Italy," p. 957.

"Restorations of a Cretaceous mosasaur named *Plotosaurus*": Hilton, *Dinosaurs and Other Mesozoic Reptiles of California,* pp. 89, 108.

"The toothed bird, *Hesperornis,* for example": Gregory, "Convergent Evolution," p. 345.

39 "In 1996, paleontologists tentatively announced": Bell et al., "The First Direct Evidence of Live Birth in Mosasauridae."

"In 2001, two paleontologists described embryos": Caldwell and Lee, "Live Birth in Cretaceous Marine Lizards (Mosasauroids)," p. 2397.

40 "Of 150,000 late Cretaceous to early Tertiary fossil specimens": Dingus and Rowe, *The Mistaken Extinction,* p. 236.

"In 2003, California's Mesozoic record": Hilton, *Dinosaurs and Other Mesozoic Reptiles of California,* p. 75.

40 "In 2005, a group of scientists judged an Antarctic Cretaceous fossil": Clarke et al., "Definitive Fossil Evidence for the Extant Avian Radiation in the Cretaceous," p. 305.

41 "At a place called Crowley's Ridge": Campbell and Lee, "Tails of *Hoffmani.*"

CHAPTER 5. COPE'S ELUSIVE OPHIDIANS

44 "'These animals,' he proclaimed": Ballou, "Strange Creatures of the Past," p. 23.

"In the 1898 *Popular Science Monthly* article": Ballou, "The Serpentlike Sea Saurians," p. 222.

"We saw some Bonetas swimming": Osborn, *Cope,* p. 43.

"I have been at the Museum": ibid., p. 40.

45 "Koch had found the bones": Wendt, *Before the Deluge,* p. 271.

"We may now look upon the mosasaurs": Cope, "On the Reptilian Orders of Pythonomorpha and Streptosauria," *Proceedings of the Boston Society of Natural History* (cited in Ellis, *Sea Dragons,* p. 210).

46 "'[I]n this country,' he announced": Marsh, "Gigantic Fossil Serpent from New Jersey," *American Naturalist* 4 (1871): 254.

"One, he avowed in a government report": Ellis, *Sea Dragons,* p. 207.

"We found the bones of the head": Osborn, *Cope,* p. 163.

47 "If I had the money": ibid., p. 517.

"Returning west in 1892": ibid., p. 442.

"After reading in a Philadelphia morning paper": ibid., p. 518.

48 "In 1997, however, Michael W. Caldwell and Michael S. Y. Lee": Caldwell and Lee, "A Snake with Legs from the Marine Cretaceous of the Middle East," p. 705.

"In another article": Lee and Caldwell, "*Adriosaurus* and the Affinities of Mosasaurs, Dolichosaurs, and Snakes," p. 926.

49 "In a 2000 article": Tchernov et al., "A Fossil Snake with Limbs," p. 210.

"In 2006, the announcement of a newly discovered Cretaceous snake": "Snake Fossil Has Two Legs," *San Francisco Chronicle,* April 20, 2006.

"Pacific rubber boas are good swimmers": Stebbins, *Field Guide to Western Reptiles and Amphibians,*" p. 142.

50 "Marsupials called stagodonts": Nick Longrich, "Aquatic Specialization in Marsupials from the Late Cretaceous of North America," in Uhen, p. 46.

CHAPTER 6. HOOVES INTO FLIPPERS

51 "When he obtained a bulbous little skull": Cope, *The Vertebrata of the Tertiary Formations of the West, Book One*, p. 247.

"The flat claws are a unique peculiarity": ibid., p. 355.

52 "The order of Cetacea": Cope, "The Cetacea," p. 599.

"In North America, the black bear": Darwin, *Origin of Species*, p. 184.

"Attempting to counter one of Owen's anti-evolution arguments": ibid., p. 304.

"Marsh, in the same 1877 speech": Marsh, "Introduction and Succession of Vertebrate Life in America," p. 373.

54 "The discovery on Vancouver Island": Kellogg, *A Review of the Archaeoceti*, p. 266.

55 "That the Sirenians are allied": Marsh, "Introduction and Succession of Vertebrate Life in America," p. 373.

56 "The derivation of the Sirenia": Cope, "The Extinct Sirenia," p. 697.

"The possible affinity of the sea-cow": Osborn, "Hunting the Ancestral Elephant in the Fayum Desert," p. 830.

57 "We would expect them": Savage et al., "Fossil Sirenia of the West Atlantic and Caribbean Region V," p. 448.

58 "Before 1970, 'the only authority'": Repenning, "Introduction," p. 301.

59 "A Missourian who'd gotten interested in marine mammal fossils": Kellogg, "The History of Whales," p. 3.

CHAPTER 7. MARSH'S DECEPTIVE DESMOSTYLIANS

60 "We may assign the relationships": Kellogg, "Pinnipeds from Miocene and Pliocene Deposits," p. 97.

61 "I wish I could describe the coast there": Brewer, *Up and Down California*, p. 130.

"The rest of the skeleton was missing": Marsh, "Notice of a New Fossil Sirenian," p. 95.

62 "Legend has it that Cope": Romer, "Cope versus Marsh," p. 202.

"We know how he acquired *Desmostylus*": Vanderhoof, "A Study of the Miocene Sirenian *Desmostylus*," p. 170.

"His diary for October 13": Ray, "Fossil Marine Mammals of Oregon," p. 424.

62 "I spent two days with Prof. Condon": Osborn, *Cope,* p. 268.

"He does not think": ibid.

"Condon had loaned Marsh a box of bones": Schuchert and LeVene, *O. C. Marsh,* p. 181.

63 "Osborn decided": Osborn, "A Remarkable New Mammal from Japan," p. 713.

"One of these came from Thomas Condon": Merriam, "Notes on the *Desmostylus* of Marsh," p. 406.

"One prominent Austrian, Othenio Abel": Vanderhoof, "A Study of the Miocene Sirenian *Desmostylus,*" p. 169.

"The characters of the skull and dentition": Merriam, "Notes on the Genus *Desmostylus* of Marsh," p. 412.

64 "No other mammal": Vanderhoof, "A Study of the Miocene Sirenian *Desmostylus,*" p. 194.

"*Cornwallius* is probably structurally ancestral to *Desmostylus*": ibid., p. 193.

"[C]ertain basic differences": Reinhart, "Summary, PhD Dissertation, June, 1952."

"Range from semi-amphibious forms": Reinhart, "Diagnosis of a New Mammalian Order, Desmostylia," p. 187.

65 "Some paleontologists suggested": Berta and Sumich, *Marine Mammals,* p. 97.

66 "On the other hand, Daryl Domning, a sirenian expert": Domning, "The Terrestrial Posture of Desmostylians," p. 99.

"And a *Desmostylus* skeleton photographed standing at Japan's Hokkaido University": Reinhart, "Summary, PhD Dissertation, June, 1952."

"Domning's statement reflected the discovery": Muizon and McDonald, "An Aquatic Sloth from the Pliocene of Peru," p. 224.

67 "As Daryl Domning showed me": Daryl Domning, personal communication, May 5, 2005.

"The tethytherian order Desmostylia": Domning, Ray, and McKenna, "Two New Oligocene Demostylians and a Discussion of Tethytherian Systematics," p. 48.

CHAPTER 8. EMLONG'S WHALE

68 "Knowledge accumulates no faster": Ray, "Fossil Marine Mammals of Oregon," p. 421.

69 "I am sure it is a great find": Domning, Ray, and McKenna, "Two New Oligocene Desmostylians and a Discussion of Tethytherian Systematics," p. 2.

"Douglas Emlong's Promethean prowess": ibid., p. 1.

"Emlong's instant intuition": ibid., p. 2.

"Emlong had started collecting": English Term Paper, Period Four, Gleneden Beach, April 19, 1959, Record Unit 7348, Emlong Papers, Smithsonian Institution Archives, p. 1.

"It would probably seem": ibid., p. 7.

"I think they invited him": Interview 9518, Oral History Project, Smithsonian Institution Archives, p. 30.

"[H]e wrote to Remington Kellogg": ibid.

"He worked by inspiration": Ray, "Obituary: Douglas Ralph Emlong," p. 45.

70 "Thunderous waves lash the coast": Zahl, "Oregon's Sidewalk on the Sea," p. 719.

"Emlong took the Zahls": ibid., p. 721.

71 "I was once in love with a haunted woman": Emlong, "Advent of Immortality," p. 55.

"Clayton Ray lamented": Ray, "Obituary: Douglas Ralph Emlong," p. 45.

"[I]t didn't seem to affect his mind": Interview 9518, Oral History Project, Smithsonian Institution Archives, p. 22.

72 "Sometimes he'd be gone all day": ibid., p. 8.

"A visitor recalled the museum as 'pretty minimal'": John Ward, personal communication, June, 2004.

"Clayton Ray described how, 'with no financial support'": Ray, "Obituary: Douglas Ralph Emlong," p. 45.

"It is my desire to clear one hundred thousand dollars": Douglas Emlong, letter of July 19, 1967 [?], Record Unit 7348, Emlong Papers, Smithsonian Institution Archives.

"Another sales pitch to a millionaire private collector": Douglas Emlong, letter of July 19, 1966, Record Unit 7348, Emlong Papers, Smithsonian Institution Archives.

"It's a matter of simple justice": *Salem Capital Journal,* September 14, 1967.

73 "I cannot explain in a rational way": Clayton Ray, personal communication, May 5, 2005.

74 "The baleen consists of a row": Darwin, *Origin of Species,* 6th ed., p. 216.

74 "In answer, it may be asked": ibid., p. 217.

75 "The structures of the mandibular rami": Cope, "The Phylogeny of the Whalebone Whales," p. 572.

"They complimented him": Interview 9518, Oral History Project, Smithsonian Institution Archives, p. 29.

76 "Emlong speculated that *Aetiocetus* 'fed upon small fish'": Emlong, "A New Archaic Cetacean from the Oligocene of Northwest Oregon," p. 49.

"No morphological obstacles": ibid., p. 1.

"The presence in late Oligocene deposits of toothed whales with broad rostra": Whitmore and Sanders, "Review of the Oligocene Cetacea," p. 319.

"Emlong excludes *Aetiocetus* from the Mysticeti": Van Valen, "Monophyly or Diphyly in the Origin of Whales," p. 39.

"Indeed, *Aetiocetus* may have had incipient baleen": Berta and Sumich, *Marine Mammals,* p. 60.

77 "[T]he stepping stones by which the evolutionist of today leads the doubting brother": Marsh, "Introduction and Succession of Vertebrate Life in America," p. 352.

78 "Ewan Fordyce, a New Zealand paleontologist": Fordyce, "*Simocetus rayi* (Odontoceti-Simocetidae, New Family)," p. 185.

"Its cranial characters are more derived": Goedert and Barnes, "The Earliest Known Odontocete," p. 148.

"My opinion": Lawrence Barnes, phone interview, June 3, 2005.

79 "[T]he bone shows a basic relationship to the coracoid in cormorants": Howard, "A New Family of Pelecaniforme Birds," p. 68.

"Storrs Olson was bowled over": Ray, letter of February 14, 1974, Record Unit 7348, Emlong Papers, Smithsonian Institution Archives.

80 "In 1977, Emlong retrieved most of a bird skeleton": Olson, "A New Genus of Penguin-like Pelecaniform Bird from the Oligocene of Washington," p. 56.

81 "Four more plotopterid species": Olson and Hasegawa, "Fossil Counterparts of Giant Penguins from the North Pacific," p. 689.

CHAPTER 9. PAWS INTO FLIPPERS

82 "It reminded me": Thomson, *The People of the Sea,* p. 198.

83 "So little was known about fossil pinnipeds": Marsh, "Vertebrate Life in America," p. 373.

"There he fulfilled Cope's ambition": Wortman, "A New Fossil Seal from the Marine Miocene of the Oregon Coast Region," p. 90.

84 "Professor Condon has kindly permitted me": ibid., p. 89.

"In the fossil seal before us": ibid., p. 90.

"He thought *Desmatophoca*'s teeth, jaws, and cranium were more like those of a creodont": ibid., p. 91.

"Naming it *Enaliarctos*": Mitchell and Tedford, "The Enaliarctinae," p. 205.

85 "The invasion of a radically new adaptive zone": ibid., p. 208.

"The area of origin": ibid., p. 279.

86 "There are few important features": ibid., p. 278.

"We recognize the possibility": ibid.

87 "It seems to me": Tedford, "Relationships of Pinnipeds to Other Carnivores (Mammalia)," p. 373.

"The shores of the North Atlantic Ocean": Ray, "Geography of Phocid Evolution," p. 403.

88 "And Tedford noted": Tedford, "Relationships of Pinnipeds to Other Carnivores (Mammalia)," p. 373.

"He made many trips": Interview 9518, Oral History Project, Smithsonian Institution Archives, p. 37.

"I will head *[sic]* your advice": Emlong, letter of April 3, 1974, Record Unit 7348, Emlong Papers, Smithsonian Institution Archives.

"He banged his head": Interview 9518, Oral History Project, Smithsonian Institution Archives, p. 23.

"Though he recovered fairly soon": ibid., p. 20.

"As it rains almost all the time": Emlong, English Term Paper, Record Unit 7348, Emlong Papers, Smithsonian Institution Archives, p. 6.

89 "Once he had to trade the hammer he'd used": Ray, "Obituary: Douglas Ralph Emlong," p. 45.

"Jennie said that almost all his childhood reading had been 'concerned with volcanoes and earthquakes and all kinds of natural disasters'": Interview 9518, Oral History Project, Smithsonian Institution Archives, p. 9.

"At first he thought it was pneumonia": ibid., p. 21.

"On April 15, 1975, he found a pinniped skeleton": Emlong, Collecting Note E75–19, Specimen Number 374272, Paleobiology Department, National Museum of Natural History.

90 "Recently, compelling osteological evidence": Berta, Ray, and Wyss, "Skeleton of the Oldest Known Pinniped, *Enaliarctos mealsi*," p. 60.

90 "A large number of pinniped skeletal specializations": ibid., p. 61.

"Although Repenning agreed that all pinnipeds probably had evolved from bearlike, not otterlike, animals": Repenning, "Technical Comments: Oldest Pinniped," p. 499.

91 "While it is true": Berta, Ray, and Wyss, "Skeleton of the Oldest Known Pinniped, *Enaliarctos mealsi,*" p. 500.

"In 1997, Irina Koretsky": Koretsky, "Pinniped Bones from the Late Oligocene of South Carolina," p. 58A.

"Berta acknowledged Koretsky's Oligocene fossil": Berta and Sumich, *Marine Mammals,* p. 43.

"In a 2001 publication, however": Berta, Deméré, and Adam, "The Role of Dispersal, Vicariance, and Phylogeny in Reconstructing the Biogeography of Pinnipeds," p. 33A.

"Phocids and otarioids": Koretsky and Barnes, "Origins and Relationships of Pinnipeds and the Concepts of Monophyly versus Diphyly," p. 69A.

"I still instinctively find diphyly more attractive": Clayton Ray, personal communication, Smithsonian, May 5, 2005.

92 "The Smithsonian paid his way": Sullivan, "Marine Fossils Great Passion in Life of Coastal Plunge Victim"; Ray, "Obituary: Douglas Ralph Emlong," p. 46.

"He also sulked": Clayton Ray, personal communication, Smithsonian, May 5, 2005.

"Emlong did find 'important fossils'": Sullivan, "Marine Fossils Great Passion in Life of Coastal Plunge Victim."

"Most were relatively small": Barnes, "Outline of Eastern North Pacific Fossil Cetacean Assemblages," p. 325.

"It had a skull over five feet long": Lawrence Barnes, phone interview, June 2, 2005.

93 "According to John Maisey, an ichthyologist": Maisey, *Discovering Fossil Fishes,* p. 108.

94 "The Pyramid Hill area has yielded far fewer baleen species": Barnes, "Outline of Eastern North Pacific Fossil Cetacean Assemblages," p. 325.

CHAPTER 10. SEA COWS AND OYSTER BEARS

97 "The phylogenetic position of *Proneotherium*": Deméré and Berta, "A Reevaluation of *Proneotherium repenningi* from the Miocene Astoria Formation of Oregon and Its Position as a Basal Odobenid," p. 307.

"This morphologic series reflects a functional change": ibid., p. 304.

98 "The large-tusked dugongs would have acted as keystone species in the ecosystem": Domning, "West Indian Tuskers," p. 73.

"At that time": Domning, "An Ecological Model for Late Tertiary Sirenian Evolution in the North Pacific Ocean," p. 355.

99 "The desmostylians probably shared the tropical grasses with [*Dioplotherium*]": ibid., p. 356.

"The bones were in a concretion": Ray, "Fossil Marine Mammals of Oregon," p. 429.

"Eight years later": Sullivan, "Marine Fossils Great Passion in Life for Coastal Plunge Victim."

100 "If he could get that fossil": Interview 9518, Oral History Project, Smithsonian Institution Archives, p. 24.

"He saw an otter": ibid., p. 47.

"It is no accident": Ray, "Fossil Marine Mammals of Oregon," p. 432.

101 "The waves are immensely fascinating": Emlong, English Term Paper, Record Unit 7348, Emlong Papers, Smithsonian Institution Archives, p. 7.

"That was part of the excitement": Interview 9518, Oral History Project, Smithsonian Institution Archives, p. 16.

"[T]he Oregon materials": Ray, "Fossil Marine Mammals of Oregon," p. 432.

"Actually, Ray considered this a good thing": Clayton Ray, personal communication, Smithsonian, May 5, 2005.

"The crushing cheek teeth": Tedford, Barnes, and Ray, "The Early Miocene Littoral Ursoid Carnivoran *Kolponomos*," p. 11.

102 The robust foot bones": ibid., p. 31.

"*Kolponomos* represents a unique adaptation": Berta and Sumich, *Marine Mammals,* p. 99.

"The few postcranial bones": Tedford, Barnes, and Ray, "The Early Miocene Littoral Ursoid Carnivoran *Kolponomos*," p. 11.

CHAPTER 11. THE LONG, WARM SUMMER

105 "A stratum near the hill's summit": Barnes, "Outline of Eastern North Pacific Fossil Cetacean Assemblages," p. 326.

106 "In 1911, F. M. Anderson": Kellogg, "Pelagic Animals from the Temblor Formation of the Kern River Region, California," p. 218.

"Remington Kellogg, who studied the bed": ibid.

106 "The deposit at Sharktooth Hill": quoted in Mitchell, "The Miocene Pinniped, *Allodesmus,*" p. 30.

"I believe that the bonebed may be more easily explained as the result of accumulation of bones": Barnes, "Miocene Desmatophocinae from California," p. 36.

"Barnes estimates that the bones could have piled up over a million years or more": Barnes et al., "Middle Miocene Marine Ecosystems and Cetacean Diversity," in Uhen, p. 6.

"'Nevertheless,' he wrote": Kellogg, "Pelagic Animals from the Temblor Formation of the Kern River Region, California," p. 220.

107 "In 1960, finding the traditional method of digging into the bed from the sides 'unsatisfactory'": Mitchell, "The Miocene Pinniped, *Allodesmus,*" p. 30.

"*Allodesmus* is a specialized otariid pinniped": ibid., p. 1.

108 "A number of features are shared": Berta and Sumich, *Marine Mammals,* p. 45.

"Phocids are shown to be the sister group": Deméré and Berta, "The Miocene Pinniped *Desmatophoca oregonesis* Condon, 1906, from the Astoria Formation, Oregon," pp. 113, 141.

"Its tooth and mandibular morphology": Barnes, "A New Fossil Pinniped (Mammalia: Otariidae) from the Middle Miocene Sharktooth Hill Bonebed, California," p. 1.

109 "It seems to have been the pinniped equivalent": Lawrence Barnes, phone interview, June 2, 2005.

"In 1931, Kellogg called Sharktooth Hill 'the largest fauna of cetacea'": Kellogg, "Pelagic Animals from the Temblor Formation of the Kern River Region, California," p. 303.

"'Such diversity,' he wrote": Barnes, "Outline of Eastern North Pacific Fossil Cetacean Assemblages," p. 327.

"Recent counts show that Sharktooth Hill had more kinds of sperm whales": Barnes et al., "Middle Miocene Marine Ecosystems and Cetacean Diversity," in Uhen, p. 6.

110 "Unusually large numbers of sharks": Kellogg, "Pelagic Animals from the Temblor Formation of the Kern River Region, California," p. 219.

"Mitchell listed nineteen chondrichthyan genera": Mitchell, "The Miocene Pinniped, *Allodesmus,*" p. 28.

"In 1957, at a flagstone quarry": Howard, "A Gigantic 'Toothed' Marine Bird from the Miocene of California," p. 1.

111 "When we are favored with the description and figures": ibid., p. 17.

"These birds are primitive": ibid., p. 22.

113 "Storrs Olson thought they might have been pelecaniformes": Storrs Olson, personal communication, May 2005.

CHAPTER 12. EMPTYING BAYS

114 "[E]ven when transgressions of warm seas": Domning, "Sirenian Evolution in the North Pacific Ocean," p. 140.

116 "Charles Repenning wrote in 1976": Repenning, "Adaptive Evolution of Sea Lions and Walruses," p. 381.

"Both upper and lower canines are enlarged": Barnes and Raschke, "*Gomphotaria pugnax,* a New Genus and Species of Late Miocene Dusignathine Otariid Pinniped from California," p. 1.

117 "Indeed, cetacean diversity may have increased": Barnes, "Outline of Eastern North Pacific Fossil Cetacean Assemblages," p. 336.

119 "This form is recognizable": Repenning, "Adaptive Evolution of Sea Lions and Walruses," p. 385.

"All later otariids": ibid.

"He named it *Rhabdofario*": Cavender and Miller, "*Smilodonichthys rastrosus,*" p. 40.

120 "In 1972, two ichthyologists complained": ibid., p. 1.

121 "Instead, the structure of its gills": ibid., p. 41.

CHAPTER 13. PUNCTUATED PINNIPEDS
AND DARWINIAN SIRENIANS

123 "According to Domning": Domning, "Sirenian Evolution in the North Pacific Ocean," p. 140.

"*Dusiren jordani,* under the pressure of recurrent Late Miocene cold episodes": ibid.

124 "This evolution is not more than further elaboration of the late Miocene adaptation to algae-eating": ibid., p. 141.

125 "Within the California-Baja California province": Domning, "An Ecological Model for Late Tertiary Sirenian Evolution in the North Pacific Ocean," p. 360.

"'Thus,' Domning concluded, 'the simplest conclusion is that the hydrodamaline [sea cow] lineage comprised a single evolving panmictic [all related] population'": ibid.

125 "A sirenian finally turned up in Oregon in the 1980s": Domning and Ray, "The Earliest Sirenian from the Eastern Pacific Ocean," p. 274.

"Sirenians might have been predicted to range north of California at that time": ibid., p. 273.

126 "A textbook description of the Japanese species": Berta and Sumich, *Marine Mammals,* p. 94.

"Between 8 and 4 million years ago": Repenning, "Adaptive Evolution of Sea Lions and Walruses," p. 382.

127 "About 8 million years ago": ibid., p. 385.

"In 1977, the skull of such an animal turned up": Repenning and Tedford, "Otarioid Seals of the Neogene," p. 66.

128 "It digs up buried clams": Berta and Sumich, *Marine Mammals,* p. 295.

CHAPTER 14. ADVENT OF AUTUMN

130 "Charles Repenning observed": Repenning, "Adaptive Evolution of Sea Lions and Walruses," p. 383.

132 "In the 1990s, I read about an apparent fossil phocid": Webb, "The Great Faunal Interchange," in Coates, p. 102.

134 "Methods of collecting and preserving": Kellogg, "Miocene Calvert Mysticetes Described by Cope," p. 103.

"A schooner load of bones": Cope, "The California Gray Whale," p. 655.

135 "It is now certain": Kellogg, "Miocene Calvert Mysticetes Described by Cope," p. 115.

"Cope later repented": Cope, "The Cetacea."

"They did not get their present tongue-twisting name": Deméré, "The Fossil Whale *Balaenoptera davidsonii* (Cope 1872)," p. 279.

136 "Southern California's Imperial Formation provides evidence": Webb, "The Great Faunal Interchange," in Coates, p. 103.

137 "According to Thomas Deméré": Deméré, "Two New Species of Fossil Walruses from the Upper Pliocene San Diego Formation, California," p. 77.

"According to Lawrence Barnes": Lawrence Barnes, phone interview, June 3, 2005.

139 "As a close associate and advisor": Ray, "Obituary: Douglas Ralph Emlong," p. 46.

"According to Ray, he also fell victim": Sullivan, "Marine Fossils Great Passion in Life of Coastal Plunge Victim."

"I am both curious and jealous": Emlong, letter of January 6, 1972, Record Unit 7348, Emlong Papers, Smithsonian Institution Archives.

"I spent five days a week for 12 years": "Big Fossil Hassle Embroils Oregon U.," *Scientific Research* (December 1967): p. 21.

"These songs are the sum": Emlong, letter of July 26, 1977, Record Unit 7348, Emlong Papers, Smithsonian Institution Archives.

"I walk for miles on the beach": Emlong, "Advent of Immortality," p. 32E.

140 "Jennie Emlong said": Interview 9518, Oral History Project, Smithsonian Institution Archives, pp. 39–40.

"Beings could manipulate DNA": Emlong, "Advent of Immortality," p. 130.

"I have gone through a harrowing period": Emlong, letter of June 4, 1973, Record Unit 7348, Emlong Papers, Smithsonian Institution Archives.

"Jennie said": Interview 9518, Oral History Project, Smithsonian Institution Archives, p. 49.

"I am recovering": Emlong, letter of October 1978, Record Unit 7348, Emlong Papers, Smithsonian Institution Archives.

"I know you will be thrilled": Emlong, letter of April 16, 1979, Record Unit 7348, Emlong Papers, Smithsonian Institution Archives.

141 "He called and said he'd found this very fine specimen": Interview 9518, Oral History Project, Smithsonian Institution Archives, p. 37.

"The trip produced no thrilling specimen": Ray, memorandum of July 7, 1980, "Last of Emlong Collection," Record Unit 7348, Emlong Papers, Smithsonian Institution Archives. p. 2.

142 "Whatever fleeting peace of mind": Ray, "Obituary: Douglas Ralph Emlong," p. 45.

CHAPTER 15. ICE AGE INVASIONS

145 "When Jeffries Wyman, the anatomist": Ray, "*Phoca wymani* and Other Tertiary Seals described from the Eastern Seaboard of North America," p. 5.

"A decade later": ibid., p. 6.

"With the 'fossil feud' in mind": Allen, *History of North American Pinnipeds,* p. 473.

"Paleontologists continued to name fossil Miocene seals *Phoca*": Ray, "Geography of Phocid Evolution," p. 392.

146 "When he became a professor, Williston proclaimed that 'the largest sea animals have been the final evolution of their respective races'": Wyss, "The Evolution of Body Size in Phocids," in Berta and Deméré, p. 72.

"In 1881, William Flower": ibid.

"In 1966, Edward Mitchell": Mitchell, "The Miocene Pinniped *Allodesmus*," p. 1.

"One paleontologist, André Wyss": Wyss, "The Evolution of Body Size in Phocids," in Berta and Deméré, p. 72.

147 "If, as Wyss and Annalisa Berta have maintained": Berta and Sumich, *Marine Mammals,* p. 45.

"The earliest apparent phocid seal fossils": Koretsky and Sanders, "Paleontology of the Late Oligocene Ashley and Chandler Bridge Formations of South Carolina, 1," p. 182.

"They found the bones 'closely comparable to the most specialized phocid'": ibid., p. 179.

"When Clayton Ray examined Jeffries Wyman's dusty Virginia Miocene seal bones": Ray, "*Phoca wymani* and Other Tertiary Seals Described from the Eastern Seaboard of North America," p. 20.

149 "Phocines responded to deteriorating climates": Ray, "Geography of Phocid Evolution," p. 402.

150 "At one site on Prince of Wales Island": Heaton and Grady, "The Late Wisconsin Vertebrate History of Prince of Wales Island, Southeast Alaska," p. 43.

"Yet harbor seal and sea lion fossils occur": ibid.

151 "Fossils show": Storrs Olson, personal communication, Smithsonian, May 6, 2005.

"Subsequent discoveries on other Channel islands": Miller, Mitchell, and Lipps, "New Light on the Flightless Goose, *Chendytes lawi*," p. 4.

"While marooned on Bering Island, Georg Steller encountered a flightless cormorant": Matthiessen, *Wildlife in America,* p. 99.

152 "Natives reported the existence . . . of a seabird": ibid., p. 100.

CHAPTER 16. HANDS INTO PADDLES

153 "Twenty-five-million-year-old fossil whales": Erickson, "Paleoecology of Crocodile and Whale Bearing Strata of Oligocene Age in North America," p. 7.

154 "In 1960 Professor Sir Alister Hardy": Morgan, "The Aquatic Hypothesis," p. 11.

155 "Hardy conceived his hypothesis": Morgan, *The Scars of Evolution,* p. 161.

"[A]s the descendants": Morgan, "Lucy's Child," p. 15.

"She maintained, however": Morgan, "The Aquatic Hypothesis," p. 13.

"She speculated": ibid., p. 11.

"Otherwise, she discounted paleontology": Morgan, *The Scars of Evolution,* p. 13.

"A well-known paleontologist expressed long-standing opinion": Kurtén, *The Age of Mammals,* p. 123.

"And Morgan chose a fatally anachronistic example": Morgan, "The Aquatic Hypothesis," p. 11.

156 "The only description I've seen of sea otters": Jordan, *The Fur Seals and Fur-Seal Islands of the North Pacific Ocean,* vol. 3, p. 215.

157 "When a Jesuit historian": Wallace, *The Monkey's Bridge,* p. 206.

"Steller, noting similarities between Siberian and Alaskan natives": Steller, *Journal of a Voyage with Bering,* p. 25.

159 "Carbon isotope analysis of the bones": Koppel, *Lost World,* p. 212.

"In the 1940s, Phil Orr": ibid., p. 246.

"In this way": ibid., p. 260.

160 "Paul Martin once joked": Wallace, *The Monkey's Bridge,* p. 211.

161 "This site is very puzzling and interesting to me": Emlong, letter of May 29, 1974, Record Unit 7348, Emlong Papers, Smithsonian Institution Archives.

"Ray thought the bones looked recent": Clayton Ray, personal communication, Smithsonian, May 5, 2005.

CHAPTER 17. PILEATED WOODPECKER'S BOAT

162 "Sebastián Cermeño, a Portuguese captain": Wagner, "The Voyage to California of Sebastián Rodriguez Cermeño in 1595," p. 13.

"These tule rafts": Kelly, "Coast Miwok," in Heizer, p. 419.

"They are not known to have eaten sea mammals": ibid., p. 416.

163 "More recently, the idea that it came from the eastern Pacific": Douglas Kennett, telephone interview, May 26, 2005.

"California has no physical suggestion of Polynesian presence": "Ancient Polynesians Visited State, Evidence Suggests," *San Francisco Chronicle,* June 20, 2005.

164 "Possible human sites": "Humans' Arrival Time Gets Another Look," *San Francisco Chronicle,* November 18, 2004.

"At least one west coast group": Miller, "Yuki, Huchnom, and Coast Yuki," in Heizer, p. 255.

"José de Acosta, the sixteenth-century Jesuit": Wallace, *The Monkey's Bridge,* p. 206.

"According to available records": Miller, "Yuki, Huchnom, and Coast Yuki," in Heizer, p. 250.

"In 1972, two anthropologists": ibid.

"Coast Yuki mythology has many marine characters": Gifford, "Coast Yuki Myths," p. 115.

165 "The events told in the night stories": ibid., p. 116.

"A brief one": ibid., p. 117.

"Another said": ibid., p. 119.

"A longer story, interestingly": ibid., pp. 146–51.

167 "They called themselves Ukoht-ontilka": Kroeber, *Handbook of the Indians of California,* p. 214.

"Kroeber wrote": ibid.

"The writings of many ethnographers": Bringhurst, *A Story as Sharp as a Knife,* p. 66.

"Visiting Alaska in 1879": Muir, *Travels in Alaska,* p. 72.

168 "Hereabouts was all saltwater": Bringhurst, *A Story as Sharp as a Knife,* pp. 224–30.

170 "There are many sea-serpent stories": LeBlond and Bousfield, *Cadborosaurus,* pp. 4–5.

"The Pawnees and Kiowas": Mayor, *Fossil Legends of the First Americans,* p. 192.

"The Achumwai": ibid., p. 149.

CHAPTER 18. THE END OF THE EARTH

172 "[T]he neerer still we came": Drake, *The World Encompassed,* p. 115.

"The narrator attributes this 'insufferable sharpnesse'": ibid., p. 117.

173 "It seemeth": ibid., p. 226.

"Some were so upset that they didn't eat": ibid., p. 131.

174 "On especially cold days"; ibid., p. 117.

"This last tale": Gilliam, *Island in Time,* p. 43.

"Venturing inland": Drake, *The World Encompassed,* p. 132.

"Some historians have cited this": Kelsey, *Sir Francis Drake,* p. 190.

"The poore birds and foules": Drake, *The World Encompassed,* p. 117.

175 "One thing we observed": ibid., p. 131.

"And it mentions islands": ibid., p. 134.

"In 1595, Sebastián Cermeño": Wagner, "Voyage to California of Sebastián Rodrigo Cermeño," p. 13.

"But, although he had plenty of time": ibid., p. 14.

"He listed 'many game animals'": Beebe and Senkewicz, *Lands of Promise and Despair,* p. 43.

"He was more interested": ibid., p. 45.

"There is evidence": Douglas Kennett, telephone interview, May 26, 2005.

176 "Of seabirds, except pelicans": Barco, *Historia Natural y Crónica de la Antigua California,* p. 51.

"Some of his descriptions were secondhand": ibid., p. 128.

"His manuscript includes a drawing": ibid., pl. 3.

"He probably made one of the first references": ibid., p. 131.

"[A]lthough some have called them beavers": ibid., p. 20.

"Also on both coasts of California": ibid., p. 21.

177 "When a woman fears a miscarriage": ibid.

"As Leonhard Stejneger, his biographer, observed": Stejneger, *Georg Wilhelm Steller,* p. 528.

"Amman certified": ibid., p. 126.

178 "Linnaeus and Johann Gmelin": ibid., p. 19.

179 "Steller decided to give up": Steller, *Journal of a Voyage with Bering,* p. 26.

"Naturalists, officers, and crew": Ford, *Where the Sea Breaks Its Back,* p. 70.

"Because I was the first": ibid., p. 71.

180 "The island is a succession of barren cliffs": Steller, *Journal of a Voyage with Bering,* p. 173.

"As for its body shape": ibid., p. 82.

"It has a chapter on *Simia marina*": Gesner, *Historia Animalium Liber III,* p. 1053.

"That picture shows a creature": ibid., p. 1054.

181 "We had not yet reached the beach": Ford, *Where the Sea Breaks Its Back,* p. 122.

"Ashore, arctic foxes": ibid., p. 123.

182 "If they have the luck to escape": ibid., p. 144.

"Altogether a beautiful and pleasing animal": ibid., p. 16.

"Every day for ten months": ibid., p. 163.

183 "While on Kamchatka": Stejneger, *Georg Wilhelm Steller,* p. 358.

"By the middle of June": ibid., p. 359.

184 "If I was asked to state": Ford, *Where the Sea Breaks Its Back,* p. 157.

"The parents love their offspring": Jordan, *The Fur Seals and Fur-Seal Islands of the North Pacific Ocean,* vol. 3, p. 203.

"These beasts are indeed terrible to look upon": ibid., p. 208.

"The reason why these beasts come hither": ibid, p. 210.

185 "But one must not suppose": ibid., p. 181.

"He described several new seabirds": Ford, *Where the Sea Breaks Its Back,* p. 156.

186 "Struck down": Stejneger, *Georg Wilhelm Steller,* p. 486.

"The work's Latin text": Steller, *Journal of a Voyage with Bering,* p. 4.

"When the ice melted": Ford, *Where the Sea Breaks Its Back,* p. 187.

CHAPTER 19. AN INDUSTRIAL INTERLUDE

187 "From the rising bubbles": Ford, *Where the Sea Breaks Its Back,* p. 188.

"Generally such a storm": ibid., p. 189.

188 "When we espied the ship": ibid., p. 195.

"There are Russians": Cook, *An Abridgement of Captain Cook's Last Voyage,* p. 278.

189 "They lie, in herds of many hundreds": ibid., p. 265.

"The sea animals seen": ibid., p. 219.

"The narrator remarked": ibid., p. 237.

"An estimated 5 million": Matthiessen, *Wildlife in America,* p. 106.

190 "They were common enough in San Francisco Bay": Ford, *Where the Sea Breaks Its Back,* p. 197.

"It is astonishing": La Pérouse, *The First French Expedition to California,* p. 78.

"Since France was unlikely to get in on the trade": ibid., p. 79.

"In 1811, Aleuts": Evens, *The Natural History of the Point Reyes Peninsula,* p. 168.

"According to one source, British whalers killed over 8,000 fur seals": Matthiessen, *Wildlife in America,* p. 102.

"Their skill in killing otters": Beebe and Senkewicz, *Lands of Promise and Despair,* p. 321.

"By the 1820s": Morwood, *Traveler in a Vanished Landscape,* p. 176.

191 "[S]o scarce are these animals": Kotzebue, *A New Voyage Round the World,* p. 42.

"Things were better in California": ibid., p. 121.

"Its value advances yearly": ibid., p. 46.

"An estimated 400,000 common murres": Evens, *The Natural History of the Point Reyes Peninsula,* p. 116.

192 "The mode of capturing them": Scammon, *The Marine Mammals of the North-Western Coast of North America,* p. 118.

"This species of whale": ibid., p. 33.

"Here the objects of pursuit": ibid., p. 266.

193 "[F]or many years, the Sea Otter hunters": ibid., p. 170.

"The herd at this time": ibid., p. 134.

194 "It is hardly necessary": ibid., p. 12.

"He justified himself": Scammon, "On the Cetaceans of the Western Coast of North America," p. 13.

"Scammon complained": Russell, *Eye of the Whale,* p. 561.

"He praised Scammon's work": Scammon, "On the Cetaceans of the Western Coast of North America," p. 13.

"He named a pilot whale": Russell, *Eye of the Whale,* p. 562.

195 "Some of his ideas": Scammon, *The Marine Mammals of the North-Western Coast of North America,* p. 176.

"Sometimes he did so": ibid., p. 79.

"He suspected sperm whales might have 'the faculty of communicating'": ibid., p. 78.

"He thought gray whales 'possessed of unusual sagacity'": ibid., p. 29.

"He noted the 'amorous antics'": ibid., p. 45.

"He thought harbor seals 'endowed with no little sagacity'": ibid., p. 166.

195 "The sea was quite smooth": ibid., p. 98.

196 "As he wrote, they were never seen in large groups": ibid., p. 97.

"His scrapbook contained a news article": Russell, *Eye of the Whale*, p. 590.

"Referring to the 'banded seal'": Scammon, *The Marine Mammals of the North-Western Coast of North America*, p. 140.

"Just before *Marine Mammals*' publication": Russell, *Eye of the Whale*, p. 593.

197 "The gray whale's coastal migration": Scammon, *The Marine Mammals of the North-Western Coast of North America*, p. 32.

"Elephant seals": ibid., p. 119.

"Even sea lions would 'soon be exterminated'": ibid., p. 139.

"By the time Scammon got to sea otters": ibid., p. 174.

"The civilized whaler seeks the hunted animal": ibid., p. 33.

198 "Guadalupe fur seals": Matthiessen, *Wildlife in America*, p. 103.

"An exhaustive search for sea otters": Ford, *Where the Sea Breaks Its Back*, p. 198.

"Pribilof Islands fur seals": Ellis, *The Empty Ocean*, p. 153; Matthiessen, *Wildlife in America*, p. 107.

"The whiskered faces of seals": Muir, *Travels in Alaska*, p. 150.

"One of the whales came close enough": Steinbeck and Ricketts, *The Log from the Sea of Cortez*, p. 233.

199 "Two considerations": Mendenhall, "Expert Testimony in the Bering Sea Controversy," p. 86.

CHAPTER 20. INTIMATIONS OF COMMUNICATION

200 "An excellent Point Reyes field guide": Evens, *The Natural History of the Point Reyes Peninsula*, p. 153.

"On the other hand, although the field guide": ibid., p. 151.

"Chatting in the 1930s with a Big Sur woman": Steinbeck and Ricketts, *The Log from the Sea of Cortez*, p. 183.

"Apparently, some biologists": Reidman, *Sea Otters*, p. 9.

"It was only when a reporter": Steinbeck and Ricketts, *The Log from the Sea of Cortez*, p. 183.

201 "A sea cow was reported": Stejneger, "How the Great Northern Sea Cow (*Rhytina*) Became Exterminated." p. 1049.

"I should regard fifteen hundred": ibid.

"In 1931, the headman": Ford, *Where the Sea Breaks Its Back,* p. 199.

"Although Guadalupe fur seals had been declared extinct": Matthiessen, *Wildlife in America,* p. 104.

"Two stations": Evens, *The Natural History of the Point Reyes Peninsula,* p. 167.

202 "In southwest Alaska, long an otter stronghold": *San Francisco Chronicle,* August 10, 2005.

203 "For unclear reasons, it has been dropping": "Decline of Hardy Alaskan Fur Seals Baffles Experts," *New York Times,* February 22, 2005.

"Since then, again for unclear reasons": Berta and Summich, *Marine Mammals,* pp. 45–46.

204 "According to another theory": Springer et al., "Sequential Megafaunal Collapse in the North Pacific Ocean," p. 12224.

"Orcas have been seen eating otters": Estes et al., "Killer Whale Predation on Sea Otters Linking Oceanic and Nearshore Ecosystems," p. 473.

"If sentiment's futility was self-evident": Mendenhall, "Expert Testimony in the Bering Sea Controversy," p. 86. Mendenhall also said: "[A]ll are agreed that the 'preservation' of the fur-seal species is important to mankind"; but qualified his qualification with the following: "[I]t is only necessary to determine *how* important or how large a proportion of mankind is concerned in this."

205 "A week ago": Scammon, *The Marine Mammals of the North-Western Coast of North America,* p. 71.

"Emlong was an impoverished introvert": Ray, letter to author of August 2, 2004.

206 "When John began to be interested": Jeffrey and Lilly, *John Lilly, So Far,* p. 104.

"He might experiment": ibid., p. 100.

"Yet the orcas in the area": ibid., p. 102.

207 "We began to have feelings": Lilly, "A Feeling of Weirdness," p. 71.

"The theory is as follows": Lilly, *The Mind of the Dolphin,* p. 55.

208 "If we 'teach' them": ibid., p. 47.

"It was a strange dream": Jeffrey and Lilly, *John Lilly, So Far,* p. 191.

209 "He was coming to the conclusion": ibid., p. 235.

"Lilly hoped": ibid., p. 236.

"By the time it was wrapped up": ibid., p. 239.

209 "In a 1988 speech": ibid., p. 267.

"In 1990, he urged society": ibid., p. 272.

"He tentatively classed gray whales": Lilly, *The Mind of the Dolphin*, p. 35.

210 "Annalisa Berta has noted": Berta, "What Is a Whale?" p. 181.

"When the parent animal is attacked": Scammon, *The Marine Mammals of the North-Western Coast of North America*, p. 32.

"[T]hey all 'bring to'": Scammon, *The Marine Mammals of the North-Western Coast of North America*, p. 77.

"These songs are much longer": Payne, *Among Whales*, p. 144.

211 "Humpback whales change their songs": ibid., p. 147.

"In 2005, the media": Associated Press, September 2, 2005.

"When in pursuit, the animal dives deeply": Scammon, *The Marine Mammals of the North-Western Coast of North America*, p. 135.

"Sea otters also occasionally ambush birds": Reidman, *Sea Otters*, p. 44.

"The words were 'strung together'": Reidman, *The Pinnipeds*, p. 331.

212 "[his] development was unremarkable": Ralls et al., "Vocalizations and Vocal Mimicry in Captive Harbor Seals, *Phoca vitulina*," p. 1051.

213 "I get the impression": Payne, *Among Whales*, p. 53.

"Since 1967 I have listened": ibid.

"I frequently get asked": ibid., p. 205.

"Whales have had their most advanced brains": ibid., p. 346.

EPILOGUE. THE OLD MAN OF THE SEA

217 "[A]lthough there is no reason to assume": Ellis, *Aquagenesis*, p. 185.

218 "In 2003, he dropped the sea mink idea": Ellis, *The Empty Ocean*, p. 128.

"The color of its fur resembles that of a fox": ibid., p. 129.

219 "Charles Sternberg, Edward Cope's assistant": Sternberg, *The Life of a Fossil Hunter*, p. 18.

220 "The ocean, they wrote, 'deep and black in the depths'": Steinbeck and Ricketts, *The Log from the Sea of Cortez*, pp. 30–33.

221 "The Unconscious Mind may have many of its roots in the bioplasma": Emlong, "Advent of Immortality," p. 4.

"The unconscious relates to the entire world": Williams, *Only Apparently Real*, p. 56.

222 "A 2005 *San Francisco Chronicle* article": "Lethal Beauty: The Barrier Debate," *San Francisco Chronicle,* November 1, 2005.

"Another *Chronicle* article": "Daring Rescue of Whale off Farallones," *San Francisco Chronicle,* December 14, 2005.

"A week after": "Massive Manatee Is Spotted in Hudson River," *New York Times,* August 7, 2006.

223 "When I asked the National Seashore archivist": Carola de Rooy, personal communication, Point Reyes National Seashore, June 21, 2005.

BIBLIOGRAPHY

Allen, Joel A. *History of North American Pinnipeds.* Washington: Government Printing Office, 1880.

Aranda-Manteca, F. J., D. P. Domning, and L.G. Barnes. "A New Middle Miocene Sirenian of the Genus *Metaxytherium* from Baja California and California: Relationships and Paleobiogeographic Implications." *Proceedings of the San Diego Historical Society* 129 (1994): 191–204.

Ballou, William Hosea. "Strange Creatures of the Past." *The Century Illustrated Magazine* 55 (November 1897): 15–23. "

———. The Serpentlike Sea Saurians." *Appleton's Popular Science Monthly* 53 (1898): 209–25.

Barco, Miguel del. *Historia Natural y Crónica de la Antigua California.* Mexico: Universidad Nacional Autonoma, 1973.

Barnes, Lawrence G. "Miocene Desmatophocinae from California." *University of California Publications in Geological Science* 89 (1972): 1–68.

———. "Outline of Eastern North Pacific Fossil Cetacean Assemblages." *Systematic Zoology* 25 (1976): 321–43.

———. "Fossil Enaliarctine Pinnipeds from Pyramid Hill, Kern County, California." *Contributions in Science of the Natural History Museum of Los Angeles County* 318 (1979): 1–41.

———. "Evolution, Taxonomy, and Antitropical Distribution of the Porpoises." *Marine Mammal Science* 1 (1985): 149–65.

———. "A New Fossil Pinniped (Mammalia: Otariidae) from the Middle Miocene Sharktooth Hill Bonebed, California." *Contributions in Science of the Natural History Museum of Los Angeles County* 396 (1988): 1–11.

———. "A New Enaliarctine Pinniped from the Astoria Formation, Oregon, and a Classification of the Otariidae." *Contributions in Science of the Natural History Museum of Los Angeles County* 403 (1989): 1–26.

———. "A New Genus of Middle Miocene Enaliarctine Pinniped from the Astoria Formation in Central Oregon." *Contributions in Science of the Natural History Museum of Los Angeles County* 431 (1992): 1–27.

Barnes, Lawrence G., Daryl Domning, and Clayton E. Ray. "Status of Studies on Fossil Marine Mammals." *Marine Mammal Science* 1 (1985): 15–53.

Barnes, Lawrence G., and James L. Goedert. "Marine Vertebrate Paleontology of the Olympic Peninsula." *Washington Geology* 24 (September 1996): 17–25.

Barnes, Lawrence G., and Kiyoharu Hirota. "Miocene Pinnipeds of the Otariid Subfamily Allodesminae in the North Pacific Ocean: Systematics and Relationships." *The Island Arc: Evolution and Biogeography of Fossil Marine Vertebrates in the Pacific Realm* 3 (December 1994): 329–60.

Barnes, Lawrence G., Masaichi Kimura, Hitoshi Furusawa, and Hiroshi Sawamura. "Classification and Distribution of Oligocene Aetiocetidae from Western North America and Japan." *The Island Arc: Evolution and Biogeography of Fossil Marine Vertebrates in the Pacific Realm* 3 (December 1994): 392–431.

Barnes, Lawrence G., and R. E. Raschke. "*Gomphotaria pugnax,* a New Genus and Species of Late Miocene Dusignathine Otariid Pinniped from California." *Contributions in Science of the Natural History Museum of Los Angeles County* 426 (1991): 1–16.

Barnett, Lincoln, et al. *The World We Live In.* New York: Time Inc., 1955.

Beebe, Rose Marie, and Robert M. Senkewicz. *Lands of Promise and Despair: Chronicles of Early California.* Berkeley: Heyday Books, 2001.

Bell, Gordon, et al. "The First Direct Evidence of Live Birth in Mosasauridae (Squamata): Exceptional Preservation in the Cretaceous Pierre Shale of South Dakota." *Journal of Vertebrate Paleontology* 16 (1996): 21A.

Berta, Annalisa. "New Specimens of the Pinnipediform *Pteronarctos* from the Miocene of Oregon." *Smithsonian Contributions to Paleobiology* 69 (1991): 1–33.

———. "What Is a Whale?" *Science* 263 (January 1994): 180–81.

Berta, Annalisa, and Thomas Deméré, eds. *Contributions in Marine Mammal Paleontology Honoring Frank C. Whitmore Jr.* Proceedings of the San Diego Society of Natural History 29. San Diego, 1994.

Berta, Annalisa, Thomas Deméré, and Peter Adam. "The Role of Dispersal, Vicariance, and Phylogeny in Reconstructing the Biogeography of Pinnipeds." *Abstracts, 61st Annual Meeting of the Society of Vertebrate Paleontology* (2001): 33A.

Berta, Annalisa, and Gary S. Morgan. "A New Sea Otter from the Late Miocene and Early Pliocene of North America." *Journal of Paleontology* 59 (1985): 809–19.

Berta, Annalisa, and Clayton E. Ray. "Skeletal Morphology and Locomoter Capabilities of the Archaic Pinniped *Enaliarctos mealsi.*" *Journal of Vertebrate Paleontology* 10 (1990): 141–57.

Berta, Annalisa, Clayton E. Ray, and André R. Wyss. "Skeleton of the Oldest Known Pinniped, *Enaliarctos mealsi.*" *Science* 244 (April 1989): 60–61.

Berta, Annalisa, and James L. Sumich. *Marine Mammals: Evolutionary Biology.* San Diego: Academic Press, 1999.

Brewer, William H. *Up and Down California in 1860–1864.* Berkeley: University of California Press, 1966.

Bringhurst, Robert. *A Story as Sharp as a Knife.* Lincoln: University of Nebraska Press, 1999.

Cadbury, Deborah. *Terrible Lizard: The First Dinosaur Hunters and the Birth of a New Science.* New York: Henry Holt, 2000.

Caldwell, M. W., and M. S. Y. Lee. "A Snake with Legs from the Marine Cretaceous of the Middle East." *Nature* 386 (1997): 705.

———. "Live Birth in Cretaceous Marine Lizards (Mosasauroids)." *Proceedings of the Royal Society of London* B 268 (2001): 2397–2401.

Campbell, Carl, and Thomas Lee. "Tails of *Hoffmanii:* Mosasaur Fossils in Tsunami Deposit at K/T Boundary of Southeast Missouri." *Journal of Vertebrate Paleontology* 21 (2001): 37A.

Cavender, Ted, and Robert Miller. "*Smilodonichthys rastrosus:* A New Pliocene Salmonid Fish from Western United States." *Bulletin No. 18 of the Museum of Natural History, University of Oregon* (March 1972).

Clarke, Julia A., Claudia P. Tambussi, Jorge I. Noriega, Gregory M. Erickson, and Richard A. Ketcham. "Definitive Fossil Evidence for the Extant Avian Radiation in the Cretaceous." *Nature* 433 (2005): 305–8.

Coates, Anthony, ed. *Central America: A Natural and Cultural History.* New Haven, Conn.: Yale University Press, 1997.

Condon, Thomas. "A New Fossil Pinniped (*Desmatophoca oregonensis*) from the Miocene of the Oregon Coast. *University of Oregon Bulletin, Supplement* 3 (1906): 1–14.

Conway Morris, Simon. "Showdown on the Burgess Shale: The Challenge." *Natural History* 107 (October 1999): 48–55.

Cook, James. *An Abridgement of Captain Cook's Last Voyage, by Captain King.* London: G. Kearsley, 1784.

Cope, Edward D. "Fossil Reptiles of New Jersey." *American Naturalist* 1 (1867): 22–30.

———. "The California Gray Whale." *American Naturalist* 13 (1879): 655.

———. "The Necks of the Sauropterygia." *American Naturalist* 13 (1879): 132.

———. *The Vertebrata of the Tertiary Formations of the West: Book One.* Washington, D.C.: Government Printing Office, 1883.

———. "The Cetacea." *American Naturalist* 24 (1890): 599–616.

———. "The Extinct Sirenia." *American Naturalist* 24 (1890): 697–702.

———. "Fourth Contribution to the Marine Fauna of the Miocene Period of the United States." *Proceedings of the American Philosophical Society* 34 (1895): 135–54.

———. "The Phylogeny of the Whalebone Whales." *American Naturalist* 29 (1895): 571–73.

Daeschler, Edward B., Neil H. Shubin, and Farish Jenkins, Jr. "A Devonian Tetrapod-Like Fish and the Evolution of the Tetrapod Body Plan." *Nature* 440 (2006): 757–63.

Darwin, Charles. *On the Origin of Species by Means of Natural Selection, A Facsimile of the First Edition.* Cambridge, Mass.: Harvard University Press, 1964.

———. *On the Origin of Species by Means of Natural Selection.* 6th ed. London: A. L. Burt, 1873.

Davidson, Jane P. "Bonehead Mistakes: The Background in Scientific Literature and Illustrations for Edward Drinker Cope's First Restoration of *Elasmosaurus platyurus.*" *Proceedings of the Academy of Natural Sciences of Philadelphia* 152 (October 2002): 215–40.

Deméré, Thomas. "The Fossil Whale *Balaenoptera davidsonii* (Cope 1872), with a Review of Other Neogene Species of *Balaenoptera.*" *Marine Mammal Science* 2 (1986): 277–98.

———. "Two New Species of Fossil Walruses from the Upper Pliocene San Diego Formation, California." In *Contributions in Marine Mammal Paleontology Honoring Frank C. Whitmore, Jr.,* edited by Annalisa Berta and Thomas Deméré, 77–98. Proceedings of the San Diego Society of Natural History 29. San Diego, 1994.

Deméré, Thomas, and Annalisa Berta. "A Reevaluation of *Proneotherium repenningi* from the Miocene Astoria Formation of Oregon and Its Position as a Basal Odobenid." *Journal of Vertebrate Paleontology* 21 (2001): 279–310.

———. "The Miocene Pinniped *Desmatophoca oregonensis* Condon, 1906, from the Astoria Formation, Oregon." In *Cenozoic Mammals of Land and Sea: Tributes to the Career of Clayton E. Ray,* 113–47. Smithsonian Contributions to Paleontology 93. Washington D.C.: Smithsonian Institution Press, 2002.

Dingus, Lowell, and Timothy Rowe. *The Mistaken Extinction: Dinosaur Evolution and the Origin of Birds.* New York: W. H. Freeman, 1997.

Domning, Daryl P. "An Ecological Model for Late Tertiary Sirenian Evolution in the North Pacific Ocean." *Systematic Zoology* 25 (1976): 352–61.

———. "Sirenian Evolution in the North Pacific Ocean." *University of California Publications: Bulletin of the Department of Geological Sciences* 118. Berkeley: University of California Press, 1978.

———. "West Indian Tuskers." *Natural History* 103 (April 1994): 72–73.

———. "The Terrestrial Posture of Desmostylians." In *Cenozoic Mammals of Land and Sea: Tributes to the Career of Clayton E. Ray,* 99–111. Smithsonian

Contributions to Paleontology 93. Washington D.C.: Smithsonian Institution Press, 2002.

Domning, Daryl P., and Hitoshi Furusawa. "Summary of Taxa and Distribution of Sirenia in the North Pacific Ocean." *The Island Arc: Evolution and Biogeography of Fossil Marine Vertebrates in the Pacific Realm* 3 (December 1994): 506–12.

Domning, Daryl P., and Clayton E. Ray. "The Earliest Sirenian from the Eastern Pacific Ocean." *Marine Mammal Science* 2 (1986): 263–76.

Domning, Daryl P., Clayton E. Ray, and Malcolm C. McKenna. "Two New Oligocene Desmostylians and a Discussion of Tethytherian Systematics." *Smithsonian Contributions to Paleontology* 59 (1986).

Drake, Sir Francis. *The World Encompassed.* London: W. S. W. Vaux Esq., 1854.

Eckert, Scott A. "Bound for Deep Water: Leatherback Turtles Can Pursue Their Prey Half a Mile Straight Down." *Natural History* 101 (March 1992): 29–35.

Edgar, Blake. "The Polynesian Connection." *Archaeology* 58 (March–April 2005): 42–45.

Ellis, Richard. *Monsters of the Sea.* New York: Alfred A. Knopf, 1994.

———. *Aquagenesis: The Origin and Evolution of Life in the Sea.* New York: Viking, 2001.

———. *The Empty Ocean.* Washington D.C.: Island Press, 2003.

———. *Sea Dragons: Predators of the Prehistoric Oceans.* Lawrence: University Press of Kansas, 2003.

Emlong, Douglas. "Advent of Immortality." Unpublished manuscript. Record Unit 7348, Emlong Papers, Smithsonian Institution Archives.

———. "A New Archaic Cetacean from the Oligocene of Northwest Oregon." *Bulletin No. 3 of the Museum of Natural History, University of Oregon* (October 1966).

Emry, Robert J., ed. *Cenozoic Mammals of Land and Sea: Tributes to the Career of Clayton E. Ray.* Washington D.C.: Smithsonian Institution Press, 2002.

Erickson, Bruce R. "Paleoecology of Crocodile and Whale Bearing Strata of Oligocene Age in North America." *Historical Biology* 4 (1990): 1–14.

Erickson, Jon. *Marine Geology: Undersea Landforms and Life Forms.* New York: Facts On File, 1996.

Estes, James, et al. "Killer Whale Predation on Sea Otters Linking Oceanic and Nearshore Ecosystems." *Science* 286 (October 1998): 473–75.

Evens, Jules G. *The Natural History of the Point Reyes Peninsula.* Point Reyes, Calif.: Point Reyes National Seashore Association, 1988.

Ford, Corey. *Where the Sea Breaks Its Back.* Boston: Little Brown, 1966.

Fordyce, R. Ewan. "*Simocetus rayi* (Odontoceti-Simocetidae, New Family): A Bizarre New Archaic Oligocene Dophin from the Eastern North Pacific." In *Cenozoic Mammals of Land and Sea: Tributes to the Career of Clayton E. Ray,*

185–222. Smithsonian Contributions to Paleontology 93. Washington D.C.: Smithsonian Institution Press, 2002.

Fordyce, R. Ewan, Lawrence Barnes, and Nobuyuki Miyazaki. "General Aspects of the Evolutionary History of Whales and Dolphins." *The Island Arc: Evolution and Biogeography of Fossil Marine Vertebrates in the Pacific Realm* 3 (December 1994): 373–91.

Forey, Peter, and Phillipe Janvier. "Evolution of the Early Vertebrates." *American Scientist* 82 (1994): 554–65.

Gesner, Konrad. *Historia Animalium Liber IIII: Qui Est De Piscium et Aquatilium Animantium Natura.* Zurich: Christoph Froshover, 1558.

Gifford, Edward W. "Coast Yuki Myths." *Journal of American Folk-Lore* 50 (1937): 115–72.

Gilliam, Harold. *Island in Time: The Point Reyes Peninsula.* San Francisco: Sierra Club Books, 1962.

Gingerich, Philip D., Neil A. Wells, Donald Russell, and S. M. Ibrahim Shah. "Origin of Whales in Epicontinental Remnant Seas: New Evidence from the Early Eocene of Pakistan." *Science* 20 (April 1983): 403–6.

Goedert, James L., and Lawrence G. Barnes. "The Earliest Known Odontocete: A Cetacean with Agorophiid Affinities from Latest Eocene to Earliest Oligocene Rocks in Washington State." *Paleontological Society Special Publications* 8 (1996): 148.

Gordon, Malcolm S., and Everett C. Olson. *Invasions of the Land: The Transitions of Organisms from Aquatic to Terrestrial Life.* New York: Columbia University Press, 1995.

Gould, Stephen Jay. *Wonderful Life: The Burgess Shale and the Nature of History.* New York: W.W. Norton, 1989.

———. "The Reversal of *Hallucigenia*." *Natural History* 101 (January 1992): 12–20.

Greene, Harry W., and David Cundall. "Limbless Tetrapods and Snakes with Legs." *Science* 287 (March 2000): 139–41.

Gregory, Joseph T. "Convergent Evolution: The Jaws of *Hesperornis* and the Mosasaurs." *Evolution* 5 (1951): 345–54.

Hagadorn, James W., and Ben Waggoner. "Ediacaran Fossils from the Southwestern Great Basin, United States." *Journal of Paleontology* 74 (2000): 349–59.

Hagadorn, James W., Ben Waggoner, et al. "Early Cambrian Ediacaran-Type Fossils from California." *Journal of Paleontology* 74 (2000): 731–40.

Heaton, Timothy, and Fred Grady. "The Late Wisconsin Vertebrate History of Prince of Wales Island, Southeast Alaska." In *Ice Age Cave Faunas of North America,* edited by Blaine W. Schubert, Jim I. Mead, and Russell W. Graham, 17–53. Bloomington: Indiana University Press; [Denver]: Denver Museum of Nature and Science, 2003.

Heizer, Robert, ed. *Handbook of North American Indians.* Vol. 8, *California.* Washington, D.C.: Smithsonian Institution, 1978.

Heuvelmans, Bernard. *In the Wake of Sea Serpents.* New York: Hill and Wang, 1968.

Hilton, Richard P. *Dinosaurs and Other Mesozoic Reptiles of California.* Berkeley: University of California Press, 2003.

Howard, Hildegarde. "A Gigantic 'Toothed' Marine Bird from the Miocene of California." *Santa Barbara Museum of Natural History Bulletin: Department of Geology* 1 (1957): 1–23.

———. "Additional Avian Records from the Miocene of Sharktooth Hill, California." *Contributions in Science of the Los Angeles County Museum of Natural History* 114 (1966): 1–11.

———. "A New Family of Pelecaniforme Birds." *The Condor* 71 (1969): 68–69.

———. "Late Miocene Marine Birds from Orange County, California." *Contributions in Science of the Los Angeles County Museum of Natural History* 290 (1978): 1–26.

———. "Fossil Birds from Tertiary Marine Beds at Oceanside, San Diego County, California, with Two New Species of the Genera *Uria* and *Cepphus*." *Contributions in Science of the Los Angeles County Museum of Natural History* 341 (1982): 1–15.

Hoyt, Erich. *Orca: The Whale Called Killer.* Ontario: Camden House, 1990.

Ichishima, Hiroto, Lawrence Barnes, Ewan Fordyce, Masaichi Kimura, and David Bohaska. "A Review of Kentriodontine Dolphins: Systematics and Biogeography." *The Island Arc: Evolution and Biogeography of Fossil Marine Vertebrates in the Pacific Realm* 3 (December 1994): 486–92.

Inuzuka, Norihisa. "Aquatic Adaptations in Desmostylians." *Historical Biology* 14 (2000): 97–113.

Inuzuka, Norihisa, Daryl Domning, and Clayton Ray. "Summary of Taxa and Morphological Adaptations of the Desmostylia." *The Island Arc: Evolution and Biogeography of Fossil Marine Vertebrates in the Pacific Realm* 3 (December 1994): 522–37.

Janvier, Phillipe, and Richard Lund. "*Hardistiella montanensis* N. Gen, et Sp. (Petromyzontida) from the Lower Carboniferous of Montana, with Remarks on the Affinities of the Lampreys." *Journal of Vertebrate Paleontology* 2 (1983): 407–13.

Jeffrey, Francis, and John Lilly. *John Lilly, So Far.* Los Angeles: Jeremy P. Tarcher, 1990.

Jordan, David Starr, ed. *The Fur Seals and Fur-Seal Islands of the North Pacific Ocean.* Vol. 3. Washington, D.C.: Government Printing Office, 1899.

Kellogg, Remington. "Pinnipeds from Miocene and Pliocene Deposits." *University of California Publications in Geology* 13 (1922): 23–132.

———. "The History of Whales: Their Adaptation to Life in the Water." PhD diss., University of California, Berkeley, 1928.

———. "Pelagic Mammals from the Temblor Formation of the Kern River Region, California." *Proceedings of the California Academy of Sciences* 19 (1931): 217–397.

———. *A Review of the Archaeoceti.* Carnegie Institution Publication no. 482. Washington, D.C.: Carnegie Institution, 1936.

———. "Miocene Calvert Mysticetes Described by Cope." *Smithsonian Bulletin* 247, pt. 5 (1968): 103–32.

Kelsey, Harry. *Sir Francis Drake: The Queen's Pirate.* New Haven, Conn.: Yale University Press, 1998.

Kennett, Douglas J. *The Island Chumash: Behavioral Ecology of a Maritime Society.* Berkeley: University of California Press, 2005.

Kershaw, Sarah. "Decline of Hardy Alaskan Fur Seal Baffles Experts." *New York Times,* February 22, 2005.

Kitson, Arthur. *Captain James Cook.* New York: E. P. Dutton, 1907.

Klimley, Peter A. *The Secret Life of Sharks.* New York: Simon and Schuster, 2003.

Kohno, Naoki, and Lawrence Barnes. "Miocene Fossil Pinnipeds of the Genera *Prototaria* and *Neotherium* in the North Pacific Ocean: Evolution, Relationships, and Distribution." *The Island Arc: Evolution and Biogeography of Fossil Marine Vertebrates in the Pacific Realm* 3 (December 1994): 285–308.

Koppel, Tom. *Lost World: Rewriting Prehistory, How New Science Is Tracing America's Ice Age Mariners.* New York: Atria Books, 2003.

Koretsky, Irina. "Pinniped Bones from the Late Oligocene of South Carolina: The Oldest Known True Seal." *Abstracts, 57th Annual Meeting of the Society of Vertebrate Paleontology* (1997): 58A.

Koretsky, Irina, and Lawrence G. Barnes. "Origins and Relationships of Pinnipeds and the Concepts of Monophyly versus Diphyly." *Abstracts, 63rd Annual Meeting of the Society of Vertebrate Paleontology* (2003): 69A.

Koretsky, Irina, and Albert E. Sanders. "Paleontology of the Late Oligocene Ashley and Chandler Bridge Formations of South Carolina, 1: Paleogene Pinniped Remains; The Oldest Known Seal." In *Cenozoic Mammals of Land and Sea: Tributes to the Career of Clayton E. Ray,* 179–83. Smithsonian Contributions to Paleontology 93. Washington D.C.: Smithsonian Institution Press, 2002.

Kotzebue, Otto von. *A New Voyage Round the World in the Years 1823–1826.* Bibliotheca Australiana no. 21. New York: Da Capo Press, 1967.

Kroeber, Alfred. *Handbook of the Indians of California.* Bureau of American Ethnology Bulletin 78. Washington, D.C.: Government Printing Office, 1925.

Kurtén, Björn. *The Age of Mammals.* New York: Columbia University Press, 1971.

Langston, Wann Jr. "Pterosaurs." *Scientific American* 244 (1981): 122–36.

Lanham, Url. *The Bone Hunters.* New York: Columbia University Press, 1973.

La Pérouse, Jean-François de Galaup. *The First French Expedition to California.* Translated and with an introduction by Charles N. Rudkin. Los Angeles: Glen Dawson, 1959.

LeBlond, Paul H., and Edward L. Bousfield. *Cadborosaurus: Survivor of the Deep.* Victoria, B.C.: Horsdal and Schubart, 1995.

Le Boeuf, Burney J. *Elephant Seals.* Pacific Grove: Boxwood Press, 1985.

Lee, M. S. Y., and M. W. Caldwell. "*Adriosaurus* and the Affinities of Mosasaurs, Dolichosaurs, and Snakes." *Journal of Paleontology* 74, no. 5 (2000): 915–37.

Ley, Willy. *The Lungfish, the Dodo, and the Unicorn: An Excursion into Romantic Zoology.* New York: Viking Press, 1948.

Lilly, John. *The Mind of the Dolphin: A Non-Human Intelligence.* Garden City, N.Y.: Doubleday, 1967.

———. "A Feeling of Weirdness." In *Mind in the Waters,* edited by Joan McIntyre, 71–77. San Francisco: Sierra Club Books, 1974.

Love, John A. *Sea Otters.* Golden, Colo.: Fulcrum Publishing, 1992.

Lund, Richard, and Cecile Poplin. "Fish Diversity of the Bear Gulch Limestone, Namurian, Lower Carboniferous of Montana, USA." *GEOBIOS* 32 (1999): 285–95.

Maisey, John G. *Discovering Fossil Fishes.* New York: Henry Holt, 1996.

Marsh, Othniel C. "Odontornithes, or Birds With Teeth." *American Naturalist* 9 (1875): 625–31.

———. "Introduction and Succession of Vertebrate Life in America." *American Journal of Science,* 3d ser., 14, nos. 79–84 (1877): 337–78.

———. "Notice of a New Fossil Sirenian, from California." *American Journal of Science* 35 (1888): 94–96.

Matthiessen, Peter. *Wildlife in America.* New York: Viking Press, 1964.

Mayor, Adrienne. *Fossil Legends of the First Americans.* Princeton, N.J.: Princeton University Press, 2005.

McCarren, Mark J. *The Scientific Contributions of Othniel Charles Marsh: Birds, Bones, and Brontotheres.* Peabody Museum Special Publication 15. New Haven, Conn.: Peabody Museum of Natural History, 1993.

McGowan, Christopher. *Dinosaurs, Spitfires, and Sea Dragons.* Cambridge, Mass.: Harvard University Press, 1991.

———. *The Dragon Seekers.* Cambridge, Mass.: Perseus Publishing, 2001.

McKee, Alexander. *The Queen's Corsair: Drake's Journey of Circumnavigation, 1577–1580.* New York: Stein and Day, 1978.

McMenamin, Mark A. S. "Ediacaran Biota from Sonora, Mexico." *Proceedings of the National Academy of Science* 93 (1996): 4990–93.

Mendenhall, T. C. "Expert Testimony in the Bering Sea Controversy." *Appleton's Popular Science Monthly* 52 (1897): 73–86.

Merriam, John C. "Notes on the Genus *Desmostylus* of Marsh." *University of Cal-*

ifornia Publications: Bulletin of the Department of Geological Sciences 6 (1911): 403–12.

Miller, Alden H. "An Auklet from the Eocene of Oregon." *University of California Publications: Bulletin of the Department of Geological Sciences* 20 (1931): 23–26.

Miller, Loye, Edward D. Mitchell, and Jere H. Lipps. "New Light on the Flightless Goose, *Chendytes lawi.*" *Contributions in Science of the Los Angeles County Museum of Natural History* 43 (1961).

Mitchell, Edward D. "A Walrus and Sea Lion from the Pliocene Purisima Formation at Santa Cruz, California, with Remarks on the Type Locality and Geologic Age of the Sea Lion, *Dusignathus santacruzensis* Kellogg." *Contributions in Science of the Los Angeles County Museum of Natural History* 56 (1962).

———. "The Miocene Pinniped *Allodesmus.*" *University of California Publications: Bulletin of the Department of Geological Sciences* 61 (1966).

Mitchell, Edward D., and Charles Repenning. "The Chronologic and Geographic Range of Desmostylians." *Contributions in Science of the Los Angeles County Museum of Natural History* 78 (1973).

Mitchell, Edward D., and Richard H. Tedford. "The Enaliarctinae: A New Group of Extinct Aquatic Carnivora and a Consideration of the Origin of the Otariidae." *Bulletin of the American Museum of Natural History* 151 (1973): 201–84.

Morgan, Elaine. "The Aquatic Hypothesis." *New Scientist* 1405 (April 12, 1984): 11–13.

———. "Lucy's Child." *New Scientist* 1540/1541 (December 25, 1987): 13–15.

———. *The Scars of Evolution.* Oxford: Oxford University Press, 1994.

Morwood, William. *Traveler in a Vanished Landscape: The Life and Times of David Douglas, Botanical Explorer.* New York: Clarkson N. Potter, 1973.

Motani, Ryosuke. "Phylogeny of the Ichthyopterygia." *Journal of Vertebrate Paleontology* 19 (1999): 473–96.

———. "Rulers of the Jurassic Seas." *Scientific American* 283 (2000): 52–59.

Motani, Ryosuke, Nachio Minoura, and Tatsuro Ando. "Ichthyosaurian Relationships Illuminated by New Primitive Skeletons from Japan." *Nature* 393 (1998): 255–57.

Muir, John. *Travels in Alaska.* Boston: Houghton Mifflin, 1998.

Muizon, C. de. "The Evolution of Feeding Adaptations of the Aquatic Sloth *Thalassocnus.*" *Journal of Vertebrate Paleontology* 24 (2004): 398–410.

Muizon, C. de, and H. G. McDonald. "An Aquatic Sloth from the Pliocene of Peru." *Nature* 375 (1995): 224–27.

Olson, Storrs L. "A New Genus of Penguin-like Pelecaniform Bird from the Oligocene of Washington." *Contributions in Science of the Los Angeles County Museum of Natural History* 330 (1980): 51–57.

Olson, Storrs L., and Alan Feduccia. "Presbyornis and the Origin of the Anseriformes." *Smithsonian Contributions to Zoology* 323 (1980).

Olson, Storrs L., and Yoshikazu Hasegawa. "Fossil Counterparts of Giant Penguins from the North Pacific." *Science* 206 (1979): 688–89.

———. "A New Genus and Two New Species of Gigantic Plotopteridae from the Oligocene of Japan." *Journal of Vertebrate Paleontology* 16 (1996): 742–51.

Orr, Robert T., and Roger C. Helm. *Marine Mammals of California.* Berkeley: University of California Press, 1989.

Osborn, Henry Fairfield. "A Remarkable New Mammal from Japan." *Science* 16 (1902): 713–14.

———. "Hunting the Ancestral Elephant in the Fayum Desert." *Century Magazine* 74, no. 6 (1907): 815–35.

———. *Cope: Master Naturalist.* Princeton, N.J.: Princeton University Press, 1931.

Padian, Kevin, and L. Chiappe. "Origin of Birds and Flight." *Scientific American* 278 (1998): 38–47.

Page, Lawrence M., and Brooks M. Burr. *A Field Guide to Freshwater Fishes: North America North of Mexico.* Boston: Houghton Mifflin, 1991.

Paine, Stephani. *The World of the Sea Otter.* San Francisco: Sierra Club Books, 1993.

Payne, Roger. *Among Whales.* New York: Charles Scribner's Sons, 1995.

Ralls, Katherine, Patricia Fiorelli, and Sheri Gish. "Vocalizations and Vocal Mimicry in Captive Harbor Seals, *Phoca vitulina.*" *Canadian Journal of Zoology* 63 (1985): 1050–56.

Ray, Clayton E. "Fossil Marine Mammals of Oregon." *Systematic Zoology* 25 (1976): 420–36.

———. "Geography of Phocid Evolution." *Systematic Zoology* 25 (1976): 391–407.

———. "*Phoca wymani* and Other Tertiary Seals Described from the Eastern Seaboard of North America." *Smithsonian Institution Contributions to Paleobiology* 28 (1976): 1–36.

———. "Obituary: Douglas Ralph Emlong, 1942–1980." *Society of Vertebrate Paleontology News Bulletin* 120 (October 1980): 45–46.

Reidman, Marianne. *The Pinnipeds: Seals, Sea Lions, and Walruses.* Berkeley: University of California Press, 1990.

———. *Sea Otters,* Monterey, Calif.: Monterey Bay Aquarium, 1990.

Reinhart, Roy H. "Diagnosis of a New Mammalian Order, Desmostylia," *Journal of Geology* 61 (1953): 187.

———. "A Review of Sirenia and Desmostylia." *University of California Publications in Geological Science* 36 (1959): 1–146.

Reinstedt, Randall A. *Mysterious Sea Monsters of California's Central Coast.* Carmel: Ghost Town Publications, 1997.

Renesto, Silvio, et al. "Nothosaurid Embryos from the Middle Triassic of Northern Italy: An Insight into the Viviparity of Nothosaurs?" *Journal of Vertebrate Paleontology* 23 (2001): 957–60.

Repenning, Charles A. "Adaptive Evolution of Sea Lions and Walruses." *Systematic Zoology* 25 (1976): 375–89.

———. "Introduction." *Systematic Zoology* 25 (1976): 301.

———. "Technical Comments: Oldest Pinniped." *Science* 248 (April 1990): 499.

Repenning, Charles A., and Richard H. Tedford. "Otarioid Seals of the Neogene." *U.S. Geological Survey Professional Paper* 992 (1977).

Romer, Alfred S. "Cope versus Marsh." *Systematic Zoology*, vol. 13, no. 4 (December 1964): 201–7.

Rudwick, Martin J. S. *The Meaning of Fossils: Episodes in the History of Palaeontology.* 2d ed. Chicago: University of Chicago Press, 1985.

Rue, Leonard Lee. *Furbearing Animals of North America.* New York: Crown Publishers, 1981.

Russell, Dick. *Eye of the Whale: Epic Passage from Baja to Siberia.* New York: Simon and Schuster, 2001.

Savage, R. J. G. "Review of Early Sirenia." *Systematic Zoology* 25 (1976): 344–50.

Savage, R. J. G., Daryl P. Domning, and J. G. M. Thewissen. "Fossil Sirenia of the West Atlantic and Caribbean Region V: The Most Primitive Sirenian, Owen 55." *Journal of Vertebrate Paleontology* 14 (1994): 427–49.

Scammon, Charles M. "On the Cetaceans of the Western Coast of North America." Edited by Edward Cope. *Proceedings of the Academy of Natural Sciences of Philadelphia* 21 (April 1869): 13–63.

———. *The Marine Mammals of the North-Western Coast of North America.* New York: Dover Publications, 1968. Facsimile of 1874 edition by John H. Carmany of San Francisco and G. P. Putnam's Sons of New York.

Scheffer, Victor B. *The Year of the Seal.* New York: Charles Scribner's Sons, 1970.

Schopf, William J. *Cradle of Life.* Princeton: Princeton University Press, 1999.

Schuchert, Charles, and Clara M. LeVene. *O. C. Marsh: Pioneer in Paleontology.* New Haven, Conn.: Yale University Press, 1940.

Shu, D. G., et al. "Lower Cambrian Vertebrates from South China." *Nature* 402 (1999): 42–46.

Spotila, J. R., R. D. Reine, et al. "Pacific Leatherback Turtles Face Extinction." *Nature* 405 (2000): 529–30.

Springer, Alan, et al. "Sequential Megafaunal Collapse in the North Pacific Ocean: An Ongoing Legacy of Industrial Whaling?" *Proceedings of the National Academy of Sciences* 100 (October 14, 2003): 12223–28.

Stebbins, G. Ledyard. *Darwin to DNA, Molecules to Humanity.* San Francisco: W. H. Freeman and Company, 1982.

Stebbins, Robert C. *Field Guide to Western Reptiles and Amphibians.* Boston: Houghton Mifflin, 1966.

Steinbeck, John, and Edward F. Ricketts. *The Log from the Sea of Cortez.* New York: Penguin Books, 1976.

Stejneger, Leonhard. "How the Great Northern Sea Cow (*Rhytina*) Became Exterminated." *American Naturalist* 21 (1887): 1047–54.

———. *Georg Wilhelm Steller: The Pioneer of Alaskan Natural History.* Cambridge, Mass.: Harvard University Press, 1936.

Steller, Georg Wilhelm. *Journal of a Voyage with Bering, 1741–1742.* Edited with an introduction by O. W. Frost. Translated by Margritt A. Engel and O. W. Frost. Stanford: Stanford University Press, 1988.

Sternberg, Charles H. *The Life of a Fossil Hunter.* San Diego: Jensen Printing Company, 1931.

Stirton, R. A. "A Marine Carnivore from the Clallam Miocene Formation, Washington; Its Correlation with Nonmarine Faunas." *University of California Publications in Geological Sciences* 36 (1960): 345–68.

Sugden, John. *Sir Francis Drake.* New York: Henry Holt, 1990.

Sullivan, Ann. "Marine Fossils Great Passion in Life of Coastal Plunge Victim." *Portland Oregonian,* July 10, 1980.

Taylor, Michael A. "Plesiosaurs—Rigging and Ballasting." *Nature* 290 (1981): 628–29.

Tchernov, E., O. Rieppel, H. Zaher, M. J. Polcyn, and L. J. Jacobs. "A Fossil Snake with Limbs." *Science* 287 (March 2000): 210–12.

Teather, Louise. *Place Names of Marin: Where Did They Come From?* San Francisco: Scottwall Associates, 1986.

Tedford, R. H. "Relationship of Pinnipeds to Other Carnivores (Mammalia)." *Systematic Zoology* 25 (1976): 363–74.

Tedford, R. H., L. G. Barnes, and C. E. Ray. "The Early Miocene Littoral Ursoid Carnivoran *Kolponomos:* Systematics and Mode of Life." *Proceedings of the San Diego Museum of Natural History* 29 (1994): 11–32.

Thomson, David. *The People of the Sea.* Cleveland: The World Publishing Company, 1965.

Thoreau, Henry David. *Cape Cod.* New York: Penguin Books, 1987.

Uhen, Mark D., ed. *Evolution of Aquatic Tetrapods: Fourth Triannual Convention Abstracts.* Cranbrook Institute of Science Miscellaneous Publications, vol. 1. Akron, Ohio: Cranbrook Institute of Science, 2005.

Vanderhoof, V. L. "A Study of the Miocene Sirenian *Desmostylus.*" *University of California Publications: Bulletin of the Department of Geological Sciences* 24 (1937): 169–262.

Van Valen, Leigh. "Monophyly or Diphyly in the Origin of Whales." *Evolution* 22 (1968): 37–41.

Voy, C. D. "Blue Whale Hunt in Drake's Bay." *American Naturalist* 14 (1880): 292–95.

Wade, Nicholas, ed. *The Science Times Book of Fossils and Evolution*. New York: Lyons Press, 1998.

Wagner, Henry R. "The Voyage to California of Sebastián Rodriguez Cermeño." *California Historical Society Quarterly* 3 (April 1924): 3–25.

Walker, Phillip L., Douglas J. Kennett, Terry L. Jones, and Robert De Long. "Archaeological Investigations of the Point Bennett Pinniped Rookery on San Miguel Island." *The Fifth California Islands Symposium,* 628–32. N.p.: U.S. Department of the Interior Minerals Management Service, Pacific OCS Region, 2000.

Wallace, David Rains. *The Monkey's Bridge: Mysteries of Evolution in Central America*. San Francisco: Sierra Club Books, 1997.

———. *The Bonehunters' Revenge: Dinosaurs, Greed, and the Greatest Scientific Feud of the Gilded Age*. Boston: Houghton Mifflin, 1999.

———. *Beasts of Eden: Walking Whales, Dawn Horses, and Other Enigmas of Mammal Evolution*. Berkeley: University of California Press, 2004.

Webb, William Edward. *Buffalo Land: An Authentic Account of the Discoveries, Adventures, and Mishaps of a Scientific and Sporting Party in the Wild West*. Cincinnati: E. Hannaford and Co., 1873.

Wendt, Herbert. *Before the Deluge: The Story of Paleontology*. New York: Doubleday, 1968.

Whitmore, Frank C., and Albert E. Sanders. "Review of the Oligocene Cetacea." *Systematic Zoology* 25 (1976): 304–20.

Williams, Paul. *Only Apparently Real: The World of Philip K. Dick*. New York: Arbor House, 1986.

Wortman, Jacob L. "A New Fossil Seal from the Marine Miocene of the Oregon Coast Region." *Science* 24 (1906): 89–92.

Zahl, Paul. "Oregon's Sidewalk on the Sea." *National Geographic* 120 (November 1961): 708–34.

Zimmer, Carl. *At the Water's Edge*. New York: Free Press, 1998.

INDEX

hydroids, xix
hyenas, 52, 53, 109
Hynerpeton, 16

Iceland, 203
Ichthyornis dispar, 26. *See also* birds, toothed
ichthyosaurs, 23, 24, 33–36, 38, 42
Ichthyosaurus, 22, 23, 31
Ichthyostega, 10, 14–18
Idaho, 119
iguanas, 20; marine, 20, 21
Illinois, 8
Imagotaria, 116, 121, 126, 127. *See also* odobenids
Imperial Formation, 132, 136
India, 40
Inia, 57
insects, 8, 9
International Whaling Commission, 201, 203
Inuzaka, N., 66
Inverness Ridge, 82
invertebrates, 8, 9
Iraq, 213
Ireland, 82
Italy, 138
Iwasaki, J., 63

jaguars, 148
Jamaica, 55, 57
Japan, 34, 63, 64, 66, 81, 125, 126, 137, 138, 179, 203
jaws: of mosasaurs, 38; of toothed birds, 38; of whales, 45, 76–78
jays, Steller's, 179
jellyfish, xix, xx, 4, 5, 9; lion's mane, xvii
Jenkins, Farish, 16
Jesuits, 176
"Joe and Rosie" (dolphins), 209
Juneau, 215
Jung, Carl, 220
Jungle Book, 219
Jurassic Period, 21, 24, 25, 27, 34, 36, 42

Kamchatka Peninsula, xxii, xxiii, 179, 180, 183, 185–88
Kansas, 24, 26, 27, 30, 31
Kayak Island, 179

Kellogg, Remington, 58–61; on cetaceans, 54, 55, 109, 134, 135; and Emlong, 69, 75; on pinnipeds, 106, 107; on Sharktooth Hill, 106, 107, 109, 110
Kelly, Isabel, xxvii
kelp, xviii, xxi, xxii, xxiv, 66, 95, 121, 123, 124, 131, 138, 165, 168, 192, 193, 201, 202; bull, 115
kelp forest, 115, 127, 131, 202, 204
Kennewick Man, 164
kentriodontids, 109
Kern County (California),
King Kong, xix
Kiowas, 170
Kipling, Rudyard, 219
Klamath Mountains, 3, 33, 81; and ediacarians, 3; and marine reptiles, 33–35
Klimley, Peter, 12
Koch, Albert, 45
Kodiak Island, xx, 188
Kolponomos, xxvi, 99–102, 113, 117, 130, 131
Koppel, Tom, 159
Koretsky, Irina, 91, 147
Kotzebue, Otto von, 191
Kroeber, Alfred, 164, 167
Kronosaurus, 39
Kuril Islands, 193
Kwagunt Formation, 2

La Jolla, 203
La Perouse, Jean Francois, 190
Lake Baikal, 148
Lake Tanganyika, 156
Lamarck, Jean Baptiste, 22
Lamarckism, 22, 29, 220, 221
lampreys, 11, 12, 17, 94, 119, 214
Lapland, 177, 191
Latimeria, 14. *See also* coelacanths
Laurasia, xxviii, 47
"law of higgledy-piggledy," 60, 126, 146. *See also* Darwinism, natural selection
Le Blond, Paul, xxiv, xxv, 170
Lee, Michael S. Y., 48, 49
Leidy, Joseph, 28, 32, 62, 134, 145, 147
lemurs, 51
Leptophoca, 147
leptospirosis, 217
Lewicki, James, 6, 10, 18

Scheffer, Victor B., xxiii, 218
Schopf, J. William, 1
scorpions, xxix, 8
Scotland, 8, 17, 18
Scott, William Berryman, 26
scurvy, xxvii, 176, 179, 182
sea anemones, 9
sea apes: Hardy's, 154–57, 218; Steller's, xix, xxv, xxvi, xxix, 205, 217–19; —Ellis on, xxiv, xxv, 217, 218; —Ford on, xxiv; —Scheffer on, xxiii, 218; —Stejneger on, xxiii; —Steller on, xx, xxi, xxiv, 154, 180, 181, 185, 205
sea cows, 56, 98; Steller's, xxii, xxiii, xxv, xxviii, 55, 113, 137, 152, 159, 160, 185, 201; —discovery of, xxii, xxviii; —evolution of, 124–26; —extinction of, xxiii, xxv, 160, 201; —Steller on, xxii, xxiii, 192, 193
sea lions, xix, xxiv, xxvii, xxviii, 83–92, 96, 97, 107, 113, 116, 119, 127, 144, 150, 153, 166, 191, 197, 198, 209, 224; California, xviii, xix; —conservation of, 202, 203; —habits of, 82, 211; —hunting of, 193, 197; Steller's, xix, 150, 159; —conservation of, 203, 204; —discovery of, 182; —habits of, 109; —hunting of, 197; —Steller on, 184, 185
Sea of Cortez, 198
sea serpents, 27, 31, 43–50, 145, 170, 220
sea urchins, 202, 203
sea-grass, 55, 57, 65, 66, 79, 98, 99, 113–16, 127
Seal Rock State Park (Oregon), 73
sealing industry, xxiii, 154, 186, 189–94, 197–99, 202
seals, xix, xxiv, 82–88, 90–92, 96, 108, 109, 132, 144–51, 155, 166, 173–75, 198, 209, 221, 223; Caspian, 147, 148; elephant, xix, xxviii, 86; —conservation of, 198, 201, 202; —evolution of, 132, 133, 138, 144, 146; —habits of, xx, 85, 104, 105, 107, 150; —hunting of, 191, 192, 197; gray, 82, 83, 86; harbor, xviii, xix, 86, 87, 211, 212, 219; —conservation of, 203, 204; —evolution of, 144–50; —habits of, 143, 195, 222, 224; harp, 145; hooded, 147; leopard, 109; monk, 132; ribbon (banded), 196

seaweeds, xxi
Sekwia, 2
Serpentum Indicum, Gesner's, 180, 181
sharks, xxi, 12, 17, 93, 110, 143, 165; basking, 106; brown, 110; bull, 110; great white, xix, xx, xxv, 93, 110; hammerhead, 12, 110; leopard, 12, 153; mako, 110; sevengill, 110; sixgill, 110; soupfin, 110
Sharktooth Hill, 105–13
Shastasaurus, 34, 35, 38, 39
shearwaters, 110
Shonisaurus, 35, 36
Shoshone Mountains, 35
Shotwell, J. A., 69, 75
Shubin, Neil, 16
Shumagin Island, 185
Siberia, xxii, 148, 157, 178, 186
Sikanni Chief River, 36
Silbert, John, xxiv
Silurian Period, 6, 7, 13, 15
Simia marina, Gesner's, 180
Simocetus, 78
Sirenia, 56, 64
sirenians, xxii, xxiii, xxv, xxviii, 55–58, 61–67, 69, 84, 98, 99, 102, 103, 113–17, 121, 123–26, 137, 152, 159, 160, 182–83, 185, 201
skates, xx, 110
Sky, John, 168, 219
Smilodonichthys (now *Oncorhynchus*), xxvi, 120–22. *See also* salmon, saber-toothed
Smith, James P., 33
Smith, Tom, xxvii, 219
Smithsonian Institution, 54, 59, 68, 72, 73, 79, 91, 101, 139, 141, 194
Smoky Hill River, 24, 28
snails, 5, 123
snakes, 43, 44, 46–50; marine, 44, 46–50; pachyophiid, 47, 48; yellow-bellied sea, 49
Snavely, Parke, 69, 75
Solutrean Culture, 163
Sonora (Mexico), 2
Sotalia, 57
South Carolina, 91, 92, 147, 153
South Dakota, 39
spiders, 58
sponges, 3, 9
Sputnik, 73, 139

Squalodon, 145
squalodonts, 117
Squamata, 43
squid, xx, 55, 81, 96, 104
stagodonts, 50
Stanford University, xxiv
Stebbins, G. Ledyard., 8
steelheads, 121
Steinbeck, John, xxvi, 198, 220
Stejneger, Leonhard, xxiii, 177; on sea
    cows, 201, on sea ape, xxiii; on Steller,
    177, 183, 186
Steller, Georg Wilhelm, xix, xxvi, 157, 177–
    88, 191, 192, 194–97, 204, 205, 210; on
    fur seals, xxiv, 183, 185; on salmon, 179;
    on sea apes, xx, xxi, xxiii–xxvi, 180, 181;
    on sea cows, xxii, 181–83, 185; on sea
    lions, xxiv, 184, 185; on sea otters, xxiv,
    179, 181–83, 185, 187
Sternberg, Charles, 30, 219
Stirton, R. A., 99
Strait of Juan de Fuca, 100
stromatolites, 1, 2
Sullivan, Ann, 141, 142
Sumich, James, 102
sunfish, xix
swamps, 6, 16–18
Swanton, John, 167
Sweden, 134, 135
swordfish, 36
sycamores, 81
Systema Naturae, 177

Taku Inlet, 215
tapirs, 148
Tapolobampo, 106
Taraval, Sigismundo, 176, 177
Tarsius, 51
Tchernov, Eitan, 49
Tedford, Richard, 84; on *Kolponomos,* 101,
    102; on pinnipeds, 84–89
tektites, 41
Teleoceras, 156
teleosaurs, 21
Temblor Sea, 106, 110, 113
teratorns, 113
terns, 151
Tethys Sea, 47, 48, 54, 56, 58

tetrapods, xxvii, 15–20
Thalassia, 98. *See also* sea-grass
Thalassoleon, 127, 135
thalattosaurs, 36, 37
Thalattosaurus, 37
Thoreau, Henry, 167
tide pools, 9
Tiktaalik, 16, 17
Tlingits, 162, 167, 198
Tomales Bay, 82
Tonsala, 80, 81. *See also* plotopterids
Travels in Alaska, 198
Triassic Period, 21, 33–36, 38, 42
Trinity Alps, 81
trolls, 153
tube-worms, 71
tuna, 36
tundra, 180
turtles, xix, 21, 23, 24, 28, 38, 39, 41, 42, 50,
    51, 55, 63, 106, 113, 160, 223; green, 95;
    hawksbill, 198; leatherback, xix, 204
Tylosaurus, 24–26, 47
tyrannosaurs, 19
Tyumen (Siberia), 186

unconscious mind, 219–22
ungulates, 51, 53–57, 60, 61, 67
United States Coast Guard, 142
United States Geological Survey, 2, 30–33
United States Navy, 208, 213
United States Revenue Marine, 194
University of Arizona, 158
University of British Columbia, xxv
University of California (Berkeley), 33, 35,
    36, 38, 40, 64
University of Oregon, 69, 72, 75, 142
Uronates, 29
Ursus maritimus, 151 *See also* bear, polar
Utah, 5
Utatsusaurus, 34

Valenictus, 130, 137. *See also* odobenids
valley fever, 89
Van Gogh, Vincent, 100
Van Valen, Leigh, 53, 76, 77
Vancouver Island, 54, 59, 170
Vanderhoof, V. L., 64
Vegavis, 40

Velella, xix
Virgin Islands, 206
Virginia, 145
Vizcaino, Sebastián, 175
volcanoes, 1, 19, 40, 41, 89, 102, 106, 136, 141, 148
vultures, turkey, xviii, 111, 112, 217, 222

Walcott, Charles D., 2, 4, 5, 10, 33
walruses, 86, 89–91, 97, 108, 113, 116–18, 126–30, 189, 191
Washington (state), 77, 78, 80
Washington, D.C., 33, 92
Weaverville Flora, 81
Webb, S. David, 132
Webb, William W., 30
Wells, H. G., 14
Western Union, 194
whale strandings, xxvii, 153, 164, 169, 182
whale-watching, xix, 210
whales, xxv, xxvii, 51–53, 57, 62, 63, 72–79, 84, 106, 107, 146, 153, 155, 156, 166, 179, 191, 201, 204, 209, 213, 223; archaic, 45, 50, 53–55, 57, 58, 66, 73–79; baleen, 73–79, 81, 94, 99, 106, 109, 117, 134, 135, 153, 210; beluga, 7, 137, 151; blue, xix, xx, 134, 135, 138, 176, 195; bowhead, 74, 163, 190; finback, 195; gray, xviii, xix, 133, 138, 149, 164, 176; —conservation of, xxviii, 197, 198, 201; —Cope on, 134, 135; —evolution of, 133, 135, 138, 149; —habits of, xviii, xx, 133, 135, 138, 149, 150, 192, 221; —hunting of, 163, 192, 193, 197, 198; —intelligence of, 195, 209, 210; —Scammon on, 192, 193, 195, 197, 198, 210; —taxonomy of, 134, 135, 195; humpback, xix, 35, 135, 195, 215, 222; —conservation of, 203; —evolution of, 117; —habits of, xx, 195, 215; —singing of, xx, 210, 211; killer, 44, 135, 137 ; pilot, 194; right, 190; rorqual, 134, 135; sperm,

xxv, 45, 59, 176, 190, 195; —intelligence of, 195, 207–10; —Lilly on, 207–10; —Scammon on, 195, 210; toothed, 73–79, 81, 92, 93, 99, 106, 109, 113, 117, 135, 137, 138, 208; with legs, 53, 54, 57, 60
whaling industry, 190–98, 201, 203, 204, 206
"White Seal, The," 219
Wilkes Expedition, 70
willets, xvi
Williston, Samuel, 31, 146
Wiwaxia, 4
wolverines, 191
wolves, 144
wood lice, 5
World War I, 59
World War II, 64
World We Live In, The, 6, 7, 10, 14, 18, 24, 43, 59
worms, 4, 5, 9
Wortman, Jacob L., 83, 84, 96, 108
Wyman, Jeffries, 45, 145, 147
Wyoming, 51, 53
Wyss, Andre, 90, 91, 108, 146, 147

Xiphactinus, 28

Yale University, 18, 25, 27, 28, 31, 61, 62, 83
Yaquina Flora, 81
Yates, L. G., 61, 62
Yoshiwara, S., 63
Yucatan, 41
Yukis, Coast, 164–67, 169
Yupiks, 163

Zahl, Paul, 70, 71, 89
Zallinger, Rudolph, 18, 19, 24–26, 28, 40, 43, 47, 52, 59
Zalophus californianus, 83, 178. *See also* sea lions, California
Zeuglodon, 45, 52

Text	11.25/13.5 Adobe Garamond
Display	Adobe Garamond / Perpetua
Compositor	BookMatters, Berkeley
Printer and binder	Sheridan Books, Inc.